INFRASTRUCTURE RISK MANAGEMENT PROCESSES

NATURAL, ACCIDENTAL, AND DELIBERATE HAZARDS

EDITED BY
Craig Taylor, Ph.D.
Erik VanMarcke, Ph.D.

ASCE Council on Disaster Risk Management
Monograph No. 1

ASCE *American Society of Civil Engineers*

Cataloging-in-Publication Data on file with the Library of Congress.

American Society of Civil Engineers
1801 Alexander Bell Drive
Reston, Virginia, 20191-4400

www.pubs.asce.org

A Note on the Cover

<u>Cover photograph description:</u> Los Angeles – Owens Valley Aqueduct damage during the 1920's "Water Wars" following a dynamite attack on a sag pipe (inverted siphon) in the Owens Valley. The pipe collapsed due to suction forces created from water flowing through the pipe following the blast. The collapsed pipe was washed down slope by the continued aqueduct flow coming from the top of slope.

The Los Angeles – Owens Valley Aqueduct was dynamited at least 15 times between 1924 and 1927 as a result of disputes between the City of Los Angeles and Owens Valley residents over land purchases, land prices, water rights, and other matters. During the "Water Wars" the Los Angeles water supply was threatened by hostile attack and domestic terrorism through acts of hostile seizures, bombings, kidnappings, threats, and propaganda. The City of Los Angeles employed the following strategies to counter the terrorist-type activities: negotiation and arbitration, rewards, public and private investigations, increased police and security, increased intelligence and communication technologies, improved transportation capabilities, improving water storage capabilities, and removing unethical financing for the hostile activities. The hostile attacks and the counter activities of the Los Angeles and Owens Valley "Water Wars" are among the precursors to the current nationwide terrorist threats and security countermeasures.

Prepared by: Craig Davis and LeVal Lund

Photo Credits: Los Angeles Department of Water and Power.

Contents

INTRODUCTION AND EXECUTIVE SUMMARY

This introduction and executive summary cover

- The origins of this monograph as a sequel to a previous monograph generated by the joint efforts of the members of two committees of the American Society of Civil Engineers (ASCE) and other volunteers (section 1.0),
- A broad overview of infrastructure risk management procedures and processes, along with a clarification as to why processes and procedures are emphasized in this document (section 2.0),
- Brief synopses of the papers included in this monograph (section 3.0), and
- A list of significant related publications (section 4.0)

1.0 The Monograph as a Sequel

For six years, two committees of ASCE have jointly worked on two monographs whose papers have undergone rigorous peer review. The committees involved are the Seismic Risk Committee of the Technical Council on Lifeline Earthquake Engineering (TCLEE) and the Risk and Vulnerability Committee of the Council on Disaster Risk Management (CDRM). Through these efforts, along with those of other volunteer contributors, the committees produced in March 2002 a monograph entitled *Acceptable Risk Processes: Lifelines and Natural Hazards* (edited by Craig Taylor and Erik VanMarcke).

This previous monograph contained mainly technical papers that evaluated procedures used in these "acceptable risk processes". Considering all the advances in probabilistic seismic hazard analysis over more than three decades, David Perkins elaborated a number of remaining issues having the effect that uncertainties may be significantly higher than the well-developed models indicate. Armen Der Kiureghian presented a paper explaining how to apply Bayesian methods to obtain seismic fragility models for electric power components. Stuart Werner and Craig Taylor presented issues arising when constructing seismic vulnerability models for transportation system components. Adam Rose deals with the complex issue of validating models to estimate higher-order economic losses.

A persistent problem is how to develop prescriptive criteria that provide guidance and goals for acceptable risk procedures. In the previous monograph, Keith Porter reviewed and evaluated life-safety criteria that are available, and Daniel Alesch, Robert Nagy, and Craig Taylor addressed available financial criteria.

Inasmuch as technical procedures do not comprise the full scope of acceptable risk processes, three additional papers covered communication, administration and regulation issues. From an owner's and then an engineer's perspective, Dick Wittkop and Bo Jensen addressed challenges in communicating risk results. Frank Lobedan, Thomas LaBasco, and Kenny Ogunfunmi discussed the administration of the major wharf embankment and strengthening program at the Port of Oakland. And Martin Eskijian, Ronald Heffron, and Thomas Dahlgren discussed the regulatory process for designing and implementing the engineering standards for marine oil terminals in the State of California.

On February 9, 2002, the two ASCE committees, along with other guests, met in Long Beach, California to discuss the status of the monograph project. Considering many previously proposed papers that had not been completed for the monograph about to be published and mindful that the events of September 11, 2001 had created wide interest in infrastructure risks, the joint committees decided to produce a sequel monograph. In addition, before the meeting, Jim Beavers, then Chair of TCLEE, had suggested that the topic of acceptable risk be a workshop topic for August 10, 2003, as part of the 6th U. S. Lifeline Earthquake Engineering Workshop and Conference (TCLEE2003).

At this workshop, held in Los Angeles on August 10, 2003, all presenters either had contributed to the first monograph or were developing papers for this second. With Erik VanMarcke and Craig Taylor as moderators, the following presentations were given:

- Richard Wittkop, on risk communication
- David Perkins, on earthquake hazard mapping procedures and their uncertainties
- Beverley Adams, on emerging post-disaster reconnaissance technologies
- Keith Porter, on criteria for assessing life-safety risks
- Stuart Werner, on vulnerability modeling for critical components
- James Moore II, on transportation-system risks from tsunamis
- Le Val Lund, on risk-reduction efforts at the Los Angeles Department of Water and Power (LADWP) water system, and
- Jane Preuss, on land use planning and electric power distribution risks from natural hazards

In addition, contributors to the first monograph and authors of the present monograph were asked to make presentations for two sessions, held on October 21, 2004, at the Baltimore ASCE Civil Engineering Conference & Exposition. These sessions covered risk management and acceptable risk for natural, accidental, and malicious threats. Moderated by Erik VanMarcke and Craig Taylor, these sessions featured presentations by:

- Jose Borrero, on costs of a tsunamis generated by a submarine landslide off Palos Verdes, California
- Yumei Wang and Amar Chaker, on geologic hazards affecting the Columbia River Transportation Corridor

- Mihail Popescu, on landslide risk assessment and treatment
- LeVal Lund, on the history of multi-hazard mitigation of the LADWP water system
- Adam Rose, on regional economic impacts of the 2001 electric power blackouts in Los Angeles, and
- Ruben Jongejan, on criteria for risk mitigation in the Netherlands.

2.0 Infrastructure Risk Management Procedures and Processes

Figure 1 provides a simplified outline of acceptable risk procedures for threats to infrastructure systems. According to this outline, an initial inquiry is raised pertaining to the system. This inquiry may arise from within the organization running the system or from the outside—such as by the press, professionals, regulators, or the general public. Technical procedures consist in defining the system of interest to respond to the inquiry, identifying significant hazards, assessing the vulnerability of system components to these hazards, and assessing system performance under conditions of hazard-induced damage.

Formal technical procedures have been shown to be critical in decision-making for many large infrastructure systems and/or key components of these systems (see Taylor et al., 1998). These often permit the evaluation of a range of risk-reduction alternatives and lead to affordable yet effective solutions to infrastructure system weaknesses. As a result, special emphasis has been placed on technical issues in both monographs.

At the same time, infrastructure risk management does not consist merely of making technical evaluations, as complex as these may be. In the first place, as researchers recognize (and may or may not explicitly state), in risk modeling there are always caveats, assumptions, and other contextual issues confining or conditioning the results provided. It is difficult to quantify, for instance, all "costs" of disasters. (See the H. John Heinz III Center for Science, Economics, and the Environment, 2000) It is likewise extremely challenging if not impossible to account for all uncertainties within complex evaluations.

In the second place, there are many occasions in which formal technical evaluations are either cost-prohibitive or else unnecessary. For instance, in local flood protection programs, experience with solutions to reduce flood risks to transportation systems has in some cases led to affordable solutions that do not require extensive detailed evaluations (written communications, Rebecca Quinn and Elliott Mittler, 9/04).

In the third place, the many dimensions of infrastructure risk-informed decision-making include political, social, ethical, administrative, environmental, security, and a host of other considerations that are typically not incorporated into formal technical evaluations. In a series of Earthquake Engineering Research Institute lectures, W. Petak (ATC-58 presentation, 2/24/03) has maintained that seismic risk results and their uses are

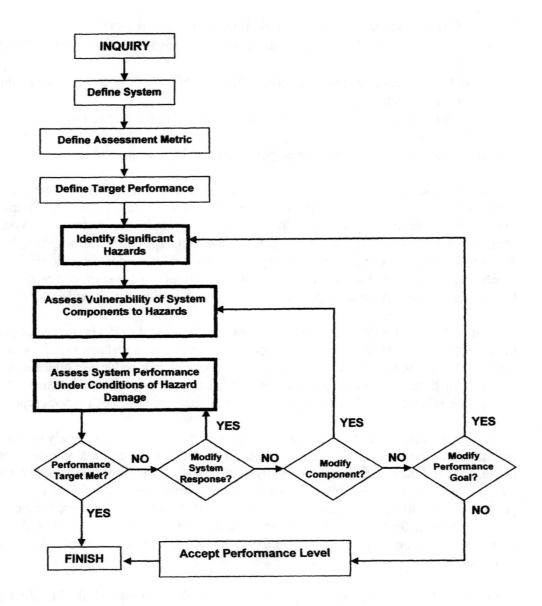

Figure 1. Decision Process for Ensuring that System Performance Goals are Met (ALA, 2004)

dependent on a variety of contexts. One cannot comprehensively explicate all contextual factors bearing on infrastructure risk management decisions. These multiple dimensions are pertinent to resolving, in Figure 1, the performance or assessment metrics to be used and the corresponding target or threshold performance levels.

In the fourth place, because there may be so many stakeholders in infrastructure risk management projects and programs, their evaluation cannot always be assumed to achieve some pre-determined level of optimality, decided solely by technical requirements. As shown in some of the papers in this monograph, the previous monograph and many case studies of infrastructure risk management programs, these programs do not usually go as planned. New information, concessions to various stakeholders, citizen rejection of proposed bonds for public works, changing directions as a result of emerging priorities such as those arising from new regulations, loss of institutional memory, changing organizational cultures, and many other factors make infrastructure risk management programs—even successful ones—dynamic. (See, for instance, James March, 1988)

3.0 Synopses of Papers in this Monograph

The first monograph covered many broad topics pertaining to acceptable risk processes for lifelines and natural hazards. However, in the early stages of development of this sequel monograph, it became clear that many important topics were in fact not treated. The first monograph's coverage focused on earthquake risks, a field that has shown quantitative sophistication for almost forty years. In spite of remaining uncertainties in estimating earthquake risks, especially to (spatially distributed) infrastructure systems, the degree of quantitative sophistication for these risks is not matched by a number of other natural hazards risks. (See American Lifelines Alliance, 2002, section 3.0). Also, accidental and malicious threats were at best an afterthought (to members of the joint ASCE committees) until September 2001.

In the previous monograph, the broad topics addressed were "Technical Issues," "Risk Criteria Issues," and "Communication, Administration and Regulation Issues." In this monograph, the broad topics covered are "Hazard Issues," "Systems Evaluation Issues," "Risk Criteria Issues," and "Systems Management Issues."

Hazard Issues.

Under the broad topic "Hazard Issues," only one paper is included. In the paper *"PSHA Uncertainty Analysis: Applications to the CEUS and the Pacific NW,"* Steven Harmsen extends a topic discussed by David Perkins in the previous monograph and in the 2003 Workshop: the magnitudes of the uncertainties in probabilistic seismic hazard analyses (PHSA's). PHSA-based estimates are used in major seismic codes and have a significant bearing on many professional, governmental, engineering, and financial activities. Most importantly, PHSA-based estimates are used in risk studies, but often without sufficient regard to (or accounting for) the uncertainty in these estimates.

This paper illustrates the quantitative sophistication in developing inputs for estimates of earthquake hazard and risk and resulting uncertainties, and presages further quantitative developments in seismic risk evaluations of infrastructure systems. Over time, the number of parties developing, reviewing, or using PHSA's and their input models has grown considerably. The USGS has produced PHSA's for over thirty years. This continuous activity has produced considerable valuable verification and validation efforts but no monopoly. Competing input models exist that express diverse opinions and results on earthquake sources, both fault-specific and random, and on how strong ground motions attenuate through rock to various sites.

For purposes of evaluating and expressing uncertainties resulting from diverse input source and attenuation models and assumptions, Harmsen, following USGS (Frankel et al., 2002), has developed a logic-tree formulation that represents the broadest features of the input alternatives at every phase. Instead of accumulating exceedance probabilities at a fixed ground motion level, however, he computes ground motions at a fixed exceedance probability level. Harmsen uses the input models and weights as found in the USGS 2002 national hazard mapping work. To supplement this USGS 2002 input information, he adds a preliminary representation of uncertainties in rates of occurrence from known faulting systems and an estimated range of uncertainty for areal source rate and b-values.

Results of these logic-tree models are expressed, for instance, in terms of probability density functions of strong ground motion values for a specific return period. These probability density functions clearly display symmetries and asymmetries, single and multiple modes, and sometimes very large ranges of uncertainty. In the main, greater uncertainties exist for sites in the Central and Eastern United States (CEUS) than in the Pacific Northwest. Alternative assumptions on major faulting sources and on attenuation models are major contributors to uncertainties. These findings can thus be used not only to guide future research but also to express more fully the range of uncertainties in earthquake hazard and risk evaluation as a result of its quantitative sophistication.

Systems Evaluation Issues.

In the paper *"The Regional Economic Cost of a Tsunami Wave Generated by a Submarine Landslide off Palos Verdes, California," Jose Borrero, Sungbin Cho, James E. Moore II, and Costas Synoloakis* discuss two topics not covered in the previous monograph: tsunamis and transportation systems analysis. According to the authors, this is the first attempt to model the prospective costs resulting from tsunamis.

The paper covers a multi-disciplinary project employing expertise in tsunamis generation and run-up analysis, transportation system analysis, and higher-order economic analysis.

The complex model consists of four main elements: a detailed assessment of the hazards caused by a submarine landslide that induces a tsunami; an account of the vulnerability of structures, especially transportation structures, in response to these hazards; a transportation network model that estimates travel-time losses and incorporates developing new ideas on how post-disaster traffic demands may be reduced by increased congestion and delays; and a regional input-output model used iteratively in conjunction with the transportation model in order to assess indirect and induced losses. Loss estimates for this hypothetical scenario range from $7B to $40B, with a range largely dependent on the how well two major ports respond to the tsunami. Estimating probabilities of this or other tsunamis still poses, however, major research challenges.

The previous monograph primarily focused on the prevention of infrastructure risks from natural hazards. Even with sound preventive measures, though, there remain residual risks, sometimes extremely large and grave. In the paper *"The Emerging Role of Remote Sensing Technology in Emergency Management," Beverley J. Adams and Charles K. Huyck* cover how remote sensing can assist not only in pre-disaster planning but in post-disaster planning as well. This emerging technology has potential applications not only for assessing landslide, earthquake, hurricane, volcano, wildfire, and other natural hazard risks, but also in coping with such anthropogenic risks as illustrated by the Columbia Space Shuttle explosion, the World Trade Center attacks, oil spills, and the movement of pollution and particulate debris.

Adams and Huyck first indicate that pre-disaster remote-sensing data are required in order to have baseline data both for loss estimation methods and for post-disaster applications. For use in response and recovery efforts, remote-sensing technology has the potential to assist in prioritizing and optimizing efforts, as well as in providing initial information that bears on the need for aid from outside the affected region. Future directions of this emerging technology include improving delivery time, training of responders and installation of this technology into emergency response networks, and more frequent baseline coverage for many regions.

Natural hazards evaluations of electric power system performance have often focused on the vulnerability of the transmission system. In the United States, this system is roughly divided into three major regions in which a single incident at one point in the system can cause outages elsewhere. Additionally, the impacts of California earthquakes on bulk substations has stood out. The focus thus has been on earthquake engineering rather than land-use management. (See ALA, 2004; Taylor et al., 1998)

In the paper *"Context and Resiliency: Influences on Electric Utility Lifeline Performance," Dorothy Reed, Jane Preuss, and Jaewook Park* focus instead on the electric power distribution system impacts of four major Pacific Northwest storms and also the 2001 Nisqually earthquake. This focus provides initial data for estimating outage times and also for assessing local vegetation management policies and practices.

Whereas damages to critical bulk substations is typically documented and analyzed in great detail, some of the data on distributions systems is incomplete or proprietary. The authors use available data to evaluate spatial and temporal extents of the five disasters. The best indicators of distribution system reliability for wind storms appears to be the ratio of the peak storm gust to the monthly gust. Outage duration models, as opposed to system restoration models, appear to be required, especially for storms that can cause damages over time even after their initial onset. Based on available evidence, the authors find that superior vegetation management policies increase distribution system reliability. Future research may confirm that such local land-use practices and policies may indeed improve the reliability of electric power systems.

Risk Criteria Issues.

In the previous monograph, two papers covered risk criteria, one pertaining to life-safety risks and the other to financial risks. In the paper *"Criteria for acceptable risk in the Netherlands,"* J. K. Vrijling, P. H. A. J. M. van Gelder, and S. J. Ouwerkerk expand this discussion by offering perspectives on life-safety risk criteria in the Netherlands.

The authors begin with a brief historical account of impacts on the Netherlands of natural hazards and of additional hazards arising from industrialization. Recent accidents, along with fairly recent developments in risk analysis, have given rise to controversies over suitable country-wide quantitative risk criteria. Those recently proposed have failed to account for risks actually assumed at the country's national airport (Schiphol), at LPG filling stations, and for roadway safety. As a result, the authors propose adding a scalar parameter covering (risk reduction) benefits and (degree of) voluntariness to the two parameters used in previous criteria, namely deaths-per-year and number of deaths in any one incident. The authors use these parameters to propose a quantitative risk criterion that accords better with practices (revealed preferences) in the Netherlands.

Systems Management Issues.

One topic not covered in the previous monograph is landslide risk management. In the paper *"Landslide Risk Assessment and Treatment," Mihail Popescu and Manoochehr Zoghi* provide a comprehensive account of the state-of-the-practice in assessing, evaluating, and managing landslide risks.

Annual estimates of landslide losses in the United States alone range from about 1.6 to 3.2 million dollars (in 2001 dollars), with 25-50 deaths (Shuster and Highland, 2001). As a consequence, significant strides are currently being made in estimating landslide risks in the United States (see Coe et al., 2004, written communication Jonathan Godt, 9/04 and David Perkins, 9/04). The Popescu and Zoghi paper draws from many regions of the world. Further strides can be made as GIS and other applications document landslides and validate and refine current landslide risk models. The authors also point out that there remain serious uncertainties in landslide risk estimates. These uncertainties are important in the overall risk management procedures outlined by these authors.

The paper *"A Preliminary Study of Geologic Hazards for the Columbia River Transportation Corridor," Yumei Wang and Amar Chaker* probes the vulnerability to multiple natural hazards in the Pacific Northwest, a region with diverse geology settings ranging from flat river valleys to steep gorge slopes. The authors examine the complex relations among different modes of transportation (highways, rail lines and river navigation) and geologic hazards, and assess their importance for the community and the region. Using data on engineered systems (bridges, roads, dams, and railroads), data on hazards (geologic hazards such as rock fall landslides, debris flow landslides, volcanic landslides, earthquake ground shaking, floods, and dam stability-related hazards) and limited economic and commerce data, they begin to assess the interdependencies and the overall vulnerability in the region, with reference to two (hypothetical) low-probability worst-case scenarios. The study results indicate that geologic hazards in the Columbia River Transportation Corridor can have a severe, long lasting impact on the economy of Oregon, affect productive capacity, and slow the pace of economic growth and development.

In the paper *"Multihazard Mitigation, Los Angeles Water System: An Historical Perspective," Le Val Lund and Craig Davis* use an historical approach to explain how the Los Angeles Department of Water and Power (LADWP) Water System has coped with earthquakes, storm water/flood, debris flow, drought, landslides, volcanism, soil settlement, cold weather, wind and seiche, hazardous materials, dam and reservoir safety, wildfire, as well as with emergency preparedness and homeland security. In its storied history over the past century, LADWP early on extended its geographic area of concern, within which this broad range of hazards had been encountered. (See Hundley, 1992, for one account of this history.)

Beginning with William Mulholland, its first Chief Engineer and General Manager, the LADWP has been a leader in many risk reduction activities. These include recognition of the need for system redundancy, early use of seismic building codes and seismic instrumentation, and use of base isolation to mitigate liquefaction-induced ground deformation effects to a water pipeline. In response to one early dam failure, and to two later dam failures, LADWP has provided lasting innovations in soil mechanics and has taken a lead in stressing dam safety evaluations and risk-reduction measures. Throughout its history, LADWP has faced accidental and malicious threats arising from such concerns as the Owens Valley controversies, the two world wars, the Cold War, and now radical Islamic terrorists.

After this survey of LADWP's comprehensive risk-reduction efforts, the authors indicate how broad management of these hazards is required. Some activities, such as compliance with the Surface Water Treatment Rule (SWTR) may not only divert resources from other activities but also potentially reduce redundant storages needed for such disasters as earthquakes. Other activities such as reducing hazardous materials impacts may contribute positively to a number of other risk-reduction efforts. Balancing these risk-reduction activities and resources needed to effect specific risk-reduction objectives requires well defined but flexible plans-of-action.

4.0 Selected Literature

American Lifelines Alliance, 2004, *Guideline for Assessing the Performance of Electric Power Systems in Natural Hazard and Human Threat Events*, supported by the Federal Emergency Management Agency through the National Institute of Building Sciences, www.americanlifelinesalliance.org.

American Lifelines Alliance, 2002, *Development of Guidelines to Define Natural Hazards Performance Objectives for Water Systems,* Vol I, September, www.americanlifelinesalliance.org

American Society of Civil Engineers, 2002, *Vulnerability and Protection of Infrastructure Systems: The State of the Art,* Reston, VA: American Society of Civil Engineers.

Beavers, James E., editor, *Advancing Mitigation Technologies and Disaster Response for Lifeline Systems: Proceedings of the Sixth U. S. Conference and workshop on Lifeline Earthquake Engineering,* Reston, VA: American Society of Civil Engineers, Technical Council on Lifeline Earthquake Engineering Monograph No. 25.

Coe, Jeffrey A., John A. Michael, Robert A. Crovelli, William Z. Savage, William T. Laprade, and William D. Nashem, 2004, "Probabilistic Assessment of Precipitation-Triggered Landslides Using Historical Records of Landslide Occurrence, Seattle, Washington," *Environmental & Engineering Geoscience*, Vol. X, No. 2, pp. 103-122, May.

Frankel, A. D., M. D. Petersen, C. S. Mueller, K. M. Haller, R. L. Wheeler, E. V. Leyendecker, R. L. Wesson, S. C. Harmsen, C. H. Cramer, D. M. Perkins, and K. S. Rukstales, 2002, *Documentation for the 2002 Update of the National Seismic Hazard Maps*, U. S. Geological Survey Open-File Report 02-420, www.usgs.gov

The H. John Heinz III Center for Science, Economics and the Environment, 2000, *The Hidden Costs of Coastal Hazards: Implications for Risk Assessment and Mitigation*, Washington, D. C.: Island Press.

Hundley, Jr., Norris, 1992, *The Great Thirst: Californias and Water, 1770s – 1990s,* Berkeley, CA: University of California Press.

March, James G, 1988, *Decisions and Organizations*, Oxford: Basil Blackwell Ltd.
Shuster, Robert L. and Lynn M. Highland, 2001, *Socioeconomic and Environmental Impacts of Landslides*, U.S.G.S. Open-File Report 01-0276, www.usgs.gov

Taylor, Craig and Erik VanMarcke, Editors, 2002, *Acceptable Risk Processes: Lifelines and Natural Hazards*, Reston, VA: American Society of Civil Engineers, Council on Disaster Reduction and Technical Council on Lifeline Earthquake Engineering, Monograph No. 21, March.

Taylor, Craig, Elliott Mittler, and LeVal Lund, 1998, *Overcoming Barriers: Lifeline Seismic Improvement Programs*, Reston, VA: American Society of Civil Engineers, Technical Council on Lifeline Earthquake Engineering Monograph No. 13, September.

HAZARD ISSUES

PSHA UNCERTAINTY ANALYSIS:
Applications to the CEUS and the Pacific NW

Stephen Harmsen
U.S. Geological Survey
Denver, CO

harmsen@usgs.gov

Abstract

Increasing reliance on probabilistic ground motion for seismic design has lead to greater demand for improved understanding and communication of seismic-hazard uncertainties. A growing sophistication in probabilistic seismic hazard analysis requires inclusion of a variety of alternative source models in many regions of the U.S. This paper examines the effect of various modeling features on ground-motion uncertainty. A logic-tree representation of the modeled epistemic uncertainty is described and used. Variability in seismic hazard is illustrated by examining hazard curves corresponding to various combinations of the model epistemic uncertainties. Fractile ground motions at various exceedance probabilities are extensively examined. We identify the PSHA model components that account for the largest amount of variability and constitute the more extreme fractiles.

The techniques that are defined are then applied to urban locations in the central and eastern U.S (CEUS) and the Pacific Northwest using the recent Frankel *et al.* (2002) PSHA model. Some carefully defined extensions to that model are included to more fully account for epistemic variability in earthquake sources. CEUS cities that are studied include two in South Carolina, Charleston and Clemson, and one in Kentucky, at Paducah. The NW sites include Seattle, Spokane, and coastal towns in Washington and Oregon.

Introduction

In recent decades, the United States Geological Survey (USGS) has published national and regional seismic hazard maps (Algermisson *et al.*, 1990; Frankel *et al.*, 1996; Wesson *et al.*, 1999; Frankel *et al.* 2002; Mueller *et al.*, 2003). These maps and associated tables exhibit mean probabilistic ground motion as a function of spectral frequency and probability of exceedance (PE) at a grid of sites that samples the U.S. and some U.S. territories. (The phrase "probability of exceedance" means, probability of one or more earthquakes with magnitude greater than a specified lower limit, at least one of which produces one or more ground-motion or oscillator pulses that exceed the specified motion.)

USGS PSHA motions, for a fixed PE, are used in the International Building Code 2000 (IBC) and other building codes (Leyendecker *et al.*, 2000). The use of probabilistic ground motion in seismic design stems from the desire, on the part of some users, for standardization and simplicity. Probabilistic ground motion provides a spatially-uniform standard over the United States of simple scalar parameters which can be used for protecting a wide category of structures from the effect of reasonably likely earthquake ground motions.

Another desire related to PSHA, one which has received increasing attention in recent years, is related to communication of model uncertainty. Such communication may require the presentation of not only the alternate input models of seismic hazard, but also effects of these variable input models on seismic-hazard estimates. How variation in input models and assumptions about their relative validity affect estimated probabilistic motions can be addressed by computing ground-motion fractiles in the distribution of seismic-hazard uncertainty. Other techniques have also been used to exhibit seismic-hazard variability. For example, Newman *et al.*, 2001, present maps of ratios of motions based on different input-model assumptions. These authors conclude that mainshock recurrence rate and attenuation model are principal components of the variability in seismic-hazard estimates in a several-state region surrounding the New Madrid Seismic Zone (NMSZ). They suggest that a high degree of uncertainty is a major feature of PSHA in that region.

USGS probabilistic seismic-hazard maps are derived from diverse models that sample the current state of the science about seismic sources and attenuation. From the input data to these maps, it is feasible to determine the probability density function (pdf) to associate with the PSHA ground-motion values for a given PE. USGS reports on seismic hazard (SH) uncertainty (e.g., Frankel *et al.*, 1997; Cramer, 2001; Petersen *et al.*, 2002) discuss some regions and cities in some detail, but for the nation as a whole, coverage is less complete. There is no single way of displaying estimates of seismic-hazard uncertainty. These reports typically describe epistemic SH model uncertainty by computing and plotting a sample coefficient of variation, which is the sample standard deviation of rate-of-exceedance estimate divided by the mean rate for a given ground motion. In distinction, this report will for the most part display ground-motion uncertainty directly, by plotting the distribution of ground-motion estimates for a given probability of exceedance.

Historically, the Nuclear Regulatory Commission has required presentation of detailed, site-specific seismic hazard uncertainty in PSHA for nuclear reactor power plant licensing, and the related literature is large (e.g., EPRI NP-4726-A). In general, however, the demand for uncertainty analysis has been limited, lagging the demand for PSHA as a uniform standard.

The determination of the input model in PSHA is a major undertaking, requiring considerable distillation of research literature, interaction with and consensus-building with expert investigators of source characteristics, propagation, and site conditions. An important underlying assumption in most USGS PSHA models is that sources have stationary rates and occur independently, i.e., there are no stress triggers or stress shadows. Alternate models, i.e., those that do not assume stationary rates, but which aim to capture the most salient features of the state of the science, are occasionally assembled for special studies. Alternate models may, for example, include time-dependent source-recurrence features, for example the Working Group on California Earthquake Probabilities (2002).

Cramer (2001) studied variability in seismic hazard at sites in a several state region around the NMSZ, considering models of source location, recurrence, size, and earthquake ground-motion attenuation that samples and more fully covers the current range of expert opinion than those included in the USGS 2002 model (Frankel *et al.*, 2002). Cramer (2001) explicitly shows the logic tree associated with his uncertainty analysis. His article may be reviewed by readers who are unfamiliar with logic trees. Another reference that provides detailed background on PSHA methods and alternative logic-tree models, including time-dependent hazard, is Thenhaus and Campbell (2002).

Logic-Tree Analysis

The logic-tree method of organizing uncertainty analysis of seismic hazard model components is used in this report. In this method, a branch is a set of components each of which is one sample of epistemic variability in the PSHA, and the suite of branches should represent the range of acceptable models. Each component of a branch includes a source model and an attenuation model. Variability of the source model may be factored into four components: (1), source size distribution, (2), source mean recurrence rate distribution, (3) distribution of source locations, and (4), other source attributes. In a logic tree, each of these source details should represent a currently plausible (preferably published) model. In this report, a logic-tree branch samples all of the independent significant sources that occur in the seismic hazard model exactly once. Alternative source distributions on a fault, e.g., distributed magnitude sources versus characteristic magnitude sources, occur on different branches. As in the USGS 2002 PSHA, we generally do not model the possibility that an identified Quaternary source is no longer active. A few exceptions to this general rule may be found in California.

Aleatory uncertainty, the part of the total uncertainty that is believed to be an intrinsic part of nature, irreducible, inevitable, could be studied in another tree - the event tree - but is not considered in this report (nor is it conventionally included in PSHA uncertainty analysis). Aleatory uncertainty is present in most input parameters of the PSHA model but is included in the logic-tree only by summation or integration. For example, we sum the exceedance rates of distributed-magnitude sources associated with a geographic grid point. In USGS PSHA, these gridded sources enter the SH model with a truncated Gutenberg-Richter (GR) magnitude-frequency distribution. An example of integrating out aleatory uncertainty is the integration of the probability density function for ground motion exceedances for a given source magnitude, source-to-site distance, and attenuation model. Summation and integration over the aleatory uncertainty means that the effects of possible but very improbable ground motions and source-model properties are present in results of the analysis, such as hazard spectra, but do not appear as distinct events as they could in an event tree.

This report assumes that all seismic and geologic model parameter estimates that enter a logic-tree analysis are mean or maximum-likelihood estimates of alternative interpretations rather than extreme fractiles of the range of each such interpretation, and that these estimates are drawn from a finite and small set of models that exhibit epistemic variability. Second, this report assumes that some combinations of parameters yield branch-models that are geologically and geophysically more consistent than others, and that it is beneficial to screen out or eliminate less defensible combinations. A large sample of source and attenuation component combinations that are internally consistent is preferred over a small sample. In this report, a given attenuation model is associated with as many sources on a logic-tree branch as possible.

Epistemic model variability has several site-dependent and site-independent features. Exploratory model analysis, for example, deaggregation (Harmsen *et al.,* 1999), pre-identifies and ranks source contributions at any given site. Overly distant sources and other marginally significant features are lumped (i.e., variability omitted) to minimize the number of epistemic branches. For example, for sites near the NMSZ, variability due to eastern seaboard sources is minimized while variability due to NMSZ source uncertainty is extensively sampled in logic-tree branches. The output at each branch's endpoint is a hazard curve, that is, an estimate of the rate of earthquakes that produce ground motion exceedances, given the components of the hazard model, and its relative weight or likelihood.

Every hazard curve in a logic-tree analysis should estimate the site's *mean* probabilistic hazard (given the validity of the model parameters in the branch). For this reason, Monte Carlo (random) sampling from continuous and discrete distribution functions for mean model parameters, is sometimes invoked to determine sample parameter settings for some or all components as they enter the logic-tree branch. In this report, Monte Carlo sampling is avoided. Monte-Carlo sampling of parameter space has the potential to yield outliers, i.e., models that may be difficult to defend as representing the mean hazard. The likelihood of generating outlier models becomes increasingly great when draws from broad-tail distributions such as the lognormal are assumed to be independent. Lognormal distributions of mean recurrence time and of maximum magnitude have been invoked to sample epistemic uncertainty (Cramer, 2001).

The First Factor in PSHA: Earthquake Sources

Earthquake sources and their occurrence rates are the major element in PSHA. Fault source contributions to site PSHA uncertainty may result from modeled interactions such as Coulomb stress transfer models, but these and other potentially plausible models are not considered. In this report, as in most of the USGS 2002 PSHA, individual faults or fault segments rupture independently. Thus, if there are N significant fault sources relative to a site, and if fault j has k_j degrees of freedom, there are potentially $\prod_{J=1}^{N} k_j$ branches before considering other sources of variation, with independent source weights. Most faults are too distant from the sites considered to impact the hazard. For this report it is assumed that a tolerable estimate of the range of variability from Quaternary fault sources is obtained by identifying the most prominent fault and lumping effects from all other fault sources, and varying properties of these two sets independently. Thus, N is at most two for sites considered in this report. Conceivably, the most prominent fault varies with PE. Activity rate, magnitude, and distance to site enter into the determination. This "binary-fault-system" assumption is believed to be valid when fault density is not too great, but should be used with caution where faults are more closely spaced. This degree of model simplification is clearly inappropriate in most of California and Nevada. Faults with segmentation models require a more elaborate organization of logic-tree branches. Segmentation models do not occur in the CEUS or Pacific NW seismic hazard models in this report or in Frankel *et al.* (2002).

Source Recurrence Times for Major Sources

An important part of PSHA uncertainty is recurrence interval for relatively large earthquakes on identified Quaternary faults. The variability of event recurrence intervals may have aleatory and epistemic components. Different investigators may disagree on how to factor rate uncertainty into these components. If in the consensus model rate variability has an epistemic component, the mean rate of hazardous sources on the fault should be varied in a logic-tree analysis.

In the input data to the USGS 2002 PSHA most CEUS source recurrence rates are fixed at one value, first, because only the mean hazard is calculated, and second, because slip-rate data are unavailable for CEUS faults. WUS fault-source recurrence rates in the USGS 2002 PSHA model vary with magnitude such that mean slip rate, as identified by geologic investigation, is conserved for each WUS fault. Thus, the range of this potentially important source of variability is not available from the input data. In this report we mildly extend the input data model by defining a fast mean rate and a slow mean rate for important sources. Fast is typically 1.2 times the rate in the USGS 2002 PSHA model and slow is typically 0.6 times the rate in that model. WUS Quaternary fault parameters such as mean recurrence interval of characteristic earthquakes are available on the web at http://geohazards.cr.usgs.gov/qfaults/index.html.

Two or more models of mean recurrence rate could result from contrasting geologic interpretation or from the assumption of two possible states of the system. Dolan *et al.* (2003) observe in the paleoseismic record an inverse correlation between activity rates for faults in the greater Los Angeles region and those in the Eastern California Seismic Zone. When one group of faults is more active the other group is less active. This two-state system implies two mean rates for each fault's activity. Each could be included in the logic tree if there is uncertainty about the current state of these interacting southern California fault systems. Other examples of two-state-of-system activity rates have been postulated for Basin and Range faults.

In this report, postulating two rates for important sources is an acknowledgment of disagreement on source rates without a comprehensive literature search to ascertain the range of ideas on this important topic. The most common reason for varying source recurrence rate is that the paleoseismic sample of recurrences for most faults is quite small, often not more than three or four intervals. Small samples suggest that mean rate estimates are generally not very secure. Small sample size is probably the main form of epistemic vulnerability in PSHA. Whereas uncertainty in mean rate for a given fault may be large if only historical and paleoseismic event information is used to inform the estimate, regional tectonic constraints are often helpful for limiting "runaway" uncertainty estimates.

Many earthquakes in the PSHA model are not associated with specific faults. For these background or "gridded" sources, USGS PSHA assumes a truncated Gutenberg-Richter (GR) distribution of earthquake magnitude. Epistemic branching results when parameters of the truncated GR distribution of source magnitude are varied systematically, that is, the *a- b-* and *Mmax* values are varied according to geophysically consistent rules. The USGS 2002 PSHA model does not contain information on epistemic variability of *a, b,* or *Mmax* . In many studies, these parameters are re-estimated after re-sampling or bootstrapping a given earthquake catalog.

To model variability in estimated rates of background seismicity, limited branching on background rates is performed here. We assume that N_k, the mean cumulative rate of earthquakes with m_b or $\mathbf{M} \geq 5$, is adequately estimated at each of the k source cells in the model of Frankel *et al.*, 2002, but that the interval rates are less secure. Although the assumption that N_k may be held fixed in the epistemic branches may seem unwarranted, its hazard consequences are the same as the assumption that the mean rate of potentially hazardous earthquakes in any annular ring about the site is adequately known.

In this report *b* is assumed to be fully correlated with *a*. Assuming complete correlation is sometimes criticized because this assumption implies that there is a magnitude for which the rate of earthquakes is known exactly, the "pivot" magnitude. Rate invariance or zero uncertainty at one magnitude on all logic-tree branches is an implausible assumption. The attractive feature (in the author's opinion) is that *b*-value may be a deterministic quantity, as yet unknown. For example, King (1983) builds a purely geometric argument that *b*=1 for many fault systems. We are sampling epistemic variability on *b*-value by varying it discretely around the values used in the USGS 2002 PSHA. In that model, *b*= 0.95 in most of the CEUS except for the Charlesvoix-St. Lawrence River region, where *b*=0.76; and *b*=0.8 in the WUS, except for deep or intraplate sources in the Puget Sound and northern California coastal regions, for which *b*=0.4. The WUS/CEUS *b*-value boundary appears in Figure 10 of Frankel *et al.* (2002).

Mmax, the maximum credible earthquake magnitude at any location, is a region-dependent parameter in the USGS 2002 PSHA. *Mmax* is kept at the USGS value in this report. *Mmax* tends to be a sensitive parameter for low-probability ground motion, such as 10^{-5}. For special studies where low-probability motion is important, such as nuclear power plant licensing studies, *Mmax* is typically varied in logic-tree branches. There is considerable mismatch between observed *Mmax* in the short historical record and model *Mmax*. In the USGS 2002 PSHA and in this report, *Mmax* is 7.5 in the extended margin of the CEUS and *Mmax* is 7.0 in the stable craton. The $\mathbf{M}7$ maximum is based on Johnston *et al.* (1994). Historical magnitudes, however, seldom much exceed $\mathbf{M}5.5$ in the vicinity of most CEUS sites. Paleoliquefaction studies often yield *Mmax* estimates that are substantially larger than those of the limited historical record. Incorporating these into PSHA models requires acceptance of several assumptions about the relation of paleoliquefaction evidence to earthquake size. In the USGS model, *Mmax* in the stable craton is determined by adding a small uncertainty to the maximum recorded $\mathbf{M}$ (6.8) in all similar domains on the planet.

Significant source zones vary, in character and number of alternatives, geographically. For example, in the southeastern U.S., the most significant contributing source zone is the Charleston S.C. characteristic source zone, with alternative fault sources and areal sources, the location and extent of which are concealed under a thick Coastal Plain sediment cover. In the USGS 2002 PSHA model, Charleston-like mainshocks are areally distributed into a very broad zone and a narrow zone of spatially uniform activity. The NMSZ characteristic earthquake zone has also been modeled as a variety of active faults, any of which may produce a dominant or modal source. Several NMSZ source characterizations, such as Cramer (2001), consider linked sequences of mainshocks. This article follows the USGS 2002 PSHA model, considering just one NMSZ characteristic source with mean recurrence time in the neighborhood of 500 years, which ruptures the entire length of the modeled fault, whose location is uncertain due to thick embayment cover, and whose size is varied in logic-tree branches.

What constitutes a significant source varies with PE: as PE decreases, the chance of a large peak ground motion from distant sources decreases. For high ground motion corresponding to low exceedance probability, only local sources may produce ground-motion exceedances with appreciable probability. The effects of ground motion attenuation and local source characterization cause the logic tree's appearance to change significantly over relatively small distances, on the order of a few hundred km or less in the CEUS, and on the order of 10s of km or less in the WUS.

The Second Factor in PSHA: Attenuation Models

Attenuation models have specific associations with sources. For example, some attenuation models, such as Youngs *et al.* (1997), are based on data from plate-subduction earthquakes, and are used only with those types of sources. Many other attenuation models, however, are associated with a wide variety of crustal earthquakes, for example, with all western U.S. shallow sources, whether they are on faults or are background (gridded) seismicity. The attenuation model of Sadigh *et al.* (1997) is unique in the USGS 2002 PSHA in that it is used to predict ground motion from subduction sources and from all shallow continental sources in the WUS.

In a logic tree analysis, one could on a single branch, randomly assign attenuation models to each successive source, or, alternatively, one could keep the same attenuation model for as many sources as possible on each branch. The latter approach is adopted here, because it makes possible the identification of the effect of attenuation functions on the resulting hazard curves. At many sites in both the CEUS and Pacific NW, attenuation models are the most important contributors to variability in the USGS 2002 PSHA model. One may compute the conditional mean and fractile hazard given attenuation model A, if each branch fixes A.

Given a magnitude and distance, ground motion from earthquakes is uncertain, and the breadth of this uncertainty, although integrated over as an aleatory uncertainty, has a considerable effect on the calculated hazard. The conventional assumption, which has some empirical basis in the WUS, is that ground-motion uncertainty has an essentially lognormal distribution. The lognormal distribution has an unbounded upper limit, physically unrealistic. In conventional PSHA, the upper part of the distribution is generally truncated, but the upper limit ground motion is an epistemic uncertainty. In the USGS 2002 PSHA model, the upper limit is defined as three sigma (σ) above the logged median, where the median and aleatory σ are functions of magnitude, distance, and possibly other fault-rupture details for each attenuation model. This 3σ epistemic limit is not varied in this report.

The Third Factor in PSHA: Epistemic Weights

For the determination of uniform hazard, weights associated with likelihood of model validity are determined through a consensus-building process and other methods. Source-model weights,

w_i, are multiplied by source rates and attenuation-model weights, w_j, are multiplied by the conditional probability of exceedance given the source and attenuation model. Equation (1) below shows how w_i and w_j are factors used to determine the mean hazard in the USGS 2002 PSHA, and other conventional PSHA.

In a logic-tree analysis, however, epistemic weights are not mathematical factors within branch hazard-curve calculations, but are external factors that determine the "weight" of the branch hazard curve, i.e., the assessment of that branch's relative validity or likelihood. In a Monte Carlo analysis, this weight governs the likelihood that a given branch will be sampled, but is otherwise not a factor. In contrast, in this report epistemic uncertainty is discretized such that all branches are included in the sample, and the weight of each branch is explicitly included in the determination of hazard-curve fractiles. The ith branch's hazard-curve weight, W_i, in the distribution of hazard curves is predetermined by multiplying the weights, w_{ij}, of the j seismic-hazard components on that branch. The normalized weight of the ith branch is its weight divided by the sum of weights of all curves in the logic-tree analysis, which is here designated U. There are typically two or more orders of magnitude in positive weights. In models considered here, many branches may have near-zero weight W_i due to distance dependence of attenuation models. Distance and other dependencies of weights may add complexity to the analysis. For example, branch weights in the Pacific NW seismic hazard model vary with SA level, requiring an SA-index k on weights.

The reported ground motion from a USGS PSHA is $SA_0 = \exp(sa_0)$, where sa_0 is the lower integration limit in the formula

$$ r = \sum_s w_s \lambda_s \sum_{a|s} w_{a|s} c \int_{sa_0}^{ul} p[sa \mid a, s] d(sa). \qquad (1a) $$

In Equation (1a), r is the rate of ground-motion exceedances at a specific site; w_s is the epistemic weight assigned to each source, s; λ_s is the mean annual frequency of occurrence of s; $a|s$ is an attenuation model used to model ground motion from s; $w_{a|s}$ is its weight in the PSHA; and the integral is the probability of exceedance of SA_0 at that site, given the source and the attenuation model. The conditional probability density function for logged ground-motion, x, is $p(x)$. Typically, and specifically for USGS PSHA, $p(x)$ is a normal, or upper-tail truncated normal density, and c is a constant that accounts for the density function's finite tail. ul is the upper limit of considered motion, which in the USGS 2002 PSHA is $\mu_A + 3\sigma_A$. Applying this upper limit in PSHA is often called 3-sigma truncation.

Notice that the exceedance rate in Equation (1a) is conditioned on a given value of sa. To get a comparable value from logic-tree component hazard curves, one chooses sa (log SA), and averages the exceedance rates r_j for all the component curves at that value of sa (Cornell, 1993). This average may be written as

$$ avg(r \mid sa) = \sum_j W_j r_j (sa) / \sum_j W_j . \qquad (1b) $$

Conventional SH analysis using component curves iterates values of *sa* until the desired exceedance probability or rate is achieved, or interpolates *sa* from bracketing sa values. Taking an average under such a condition makes sense if we have a structural design value for *sa* in mind, and we want to estimate its mean exceedance rate. *sa* may be a deterministic ground motion for some credible design scenario earthquake. In this paper the mean motion calculated from equation (1b) is reported with a "V" after it, to indicate vertical averaging.

On the other hand, the average of SA-values, given an exceedance rate, is useful for understanding the distribution of *SA* that results from alternative epistemic models. Model information at fixed *r* helps us to understand how well-constrained the mean *SA* is at a given exceedance probability. Epistemic components that produce the extreme *SA* and other fractiles at a fixed r may be examined to evaluate initial assumptions such as independence of weights.

Because the latter inquiry is the focus of this paper, the mean probabilistic motion from a logic-tree analysis reported here is given by

$$avg(SA \mid r) = \sum_j W_j SA_j(r) / \sum_j W_j \qquad (2)$$

where W_j is the product of the epistemic weights for all of the sources and attenuation models that appear in branch *j* of the logic tree. SA_j is the probabilistic motion (units typically *g*, standard gravity) determined from the set of sources and attenuation models on branch *j* for a given rate of exceedance *r*. While a given source-attenuation model pair occurs at most once in equation (1a), it occurs many times in equation (2). In this paper the two values will be compared. The approximate equality of *avg(SA)* and SA_0 in the examples below is somewhat remarkable given that equation (1) accumulates the rates exceedances of all sources at a fixed ground-motion level, and equation (2) computes exceedances at a fixed rate or probability, and therefore at a potentially different ground motion for each branch in the logic tree. Cornell (1993) states that the two mean estimates will be different. We will encounter an instance where the difference between the equation (1) and (2) estimates is more than a few percent. We identify the reason for this possibly significant difference in the section on central and non-central tendencies.

Hazard-Curve Fractiles and Epistemic Weights

One may sort hazard-curve ground motions at a specific PE, say the 2% in 50 year PE, for each of the branches of the logic tree. For the *i*th curve after sorting, if the sum of weights $\sum_{j=1}^{i-1} W_j = S/U$, then the *i*th curve is just above the *S* fractile for that PE. Because neighboring hazard curves associated with logic-tree branches may cross one or more times, any particular branch at another PE may occur at a different fractile.

The alternative way of sorting seismic hazard curves is to sort PE at a fixed ground motion. If that ground motion is the "mean" motion for that PE, i.e., the motion estimate of Equation (1a) above, then the *S* fractile curve as defined in the previous paragraph is in practice (i.e., with interpolation error) very close to and is theoretically identical to the S fractile curve as defined by the second sort (Cornell, 1993). This invariance was confirmed for the models considered in this report over PE ranging from 10% in 50 years to 1% in 1000 years. A limit to this invariance may exist. At high ground motion, H, some curves are at zero probability (or undefined in logarithmic space), because for attenuation model A, H is greater than the upper ground-motion limit of Equation 1b for all sources on the branch, in which case this branch is in the 0-probability fractile. The 0-probability fractile is here designated $S_0(H)$. However, this same branch when sorted at the PSHA model's rate, S_0, always occurs at a well-defined fractile. In this sense, sorting hazard curves at a fixed rate can be more informative than sorting them at a fixed ground motion.

The $S_0(H)$ bin, that is, the subset of SH models that cannot produce an exceedance of higher ground motion H, is not conventionally reported, although it would seem helpful to know that, say, 30% or more of the models indicate 0 likelihood of H. $k\sigma$ truncation yields lower peak ground acceleration than $(k+1)\sigma$ truncation (Bommer et al., 2004), and choices for k affect the $S_0(H)$ bin in logic-tree studies. Being able to place a physics-based upper limit on ground motion from a given earthquake for specific site conditions would be a helpful constraint on SH model uncertainty, in addition to or in some instances, in lieu of such formula truncation. Historical predictions of peak ground motion have repeatedly required upward revision (Bommer et al., 2004). An area of predominantly empirical research relative to upper ground-motion constraints is the study of precariously balanced rocks in earthquake environments, e.g., Brune (2000). In all such studies, it is important to know whether available data are a reasonably unbiased sample of the population one is trying to describe.

The identification of PSHA components that compose the median (or any other fractile) curve changes with SA level. When "the" median hazard curve is being discussed without reference to specific PE level, it should be understood that it is a hybrid curve in which different branch hazard curves have been grafted at join points between sampled ground motions.

Are w_i and w_j Correlated?

In each logic-tree branch, we would like to assign a weight to a given source-attenuation model pair that represents our belief or assessment that these models are jointly valid. Equation (1) above implicitly makes the assumption that the product of w_I and w_j represents this joint probability, i.e., the probability that the two models are simultaneously valid. Historically, many efforts in probabilistic seismic hazard estimation have been organized by identifying groups of seismologists who will work on the attenuation model problem and distinct groups of seismologists and geologists who will work on the source characterization problem. This partitioning of the work almost implies that the two efforts are distinct and that models can be combined under the assumption of model independence.

The assumption that source model and attenuation model estimation can be independent efforts, and that w_I and w_j may be multiplied in equations like (1) above, is a strong assumption about our understanding of earthquakes. Most seismologists who work with strong-motion data know that data sets are rarely available that allow "clean" separation of source from generalized attenuation effects. Generalized attenuation here means the combined effect of geometric spreading, anelastic attenuation, and seismic wave-scattering. Many data sets can be interpreted in multiple ways; in particular, correspondingly good data fits can be achieved by increasing the estimates of earthquake size and rate of generalized attenuation, or by decreasing the estimates of earthquake size and rate of attenuation. For many historical earthquakes, the data consist only of regional seismic intensity data. In this instance, the tradeoff between source size and attenuation model can be especially strong.

In this report we accept the assumption of source model and attenuation model independence with reluctance. We believe that a promising approach to model-variability reduction may be to make realistic assessments of the correlation of source properties with attenuation models. Especially in the CEUS, where most NMSZ and Charleston mainshock data are limited to intensity reports, or anywhere where modern instrumental data are unavailable for aiding in source estimation, it may be desirable from various perspectives to impose a correlation between earthquake-size models and seismic-wave attenuation models in the estimation of mean hazard and in the determination of weights for logic-tree branch hazard. In addition to above correlations, upper-bound ground motions may be correlated with specific source features, such as rupture directivity towards the site, and site features, such as subsurface focusing (Bommer *et al.*, 2004), which argues against using the same upper bound across all models.

Realized PSHA Models and Conceptual Models

Epistemic uncertainty in the description of earthquake hazard is contained in the set of conceptual geophysical, geologic and site-specific, or geotechnical, models available both in professional publications and in ongoing investigations. The portion of the conceptual models that are included in a realized PSHA model is the set having $w_1 > 0$. When performing an epistemic uncertainty analysis, one may restrict the analysis to components that appear in the realized model, or one may include models that are not part of the realized model. In the former case, the branches that represent specific epistemic model combinations are assigned weights that, as mentioned above, may be the same as those used in the realized PSHA. One could reweight components, diverging from the realized PSHA weights. While the resulting variability in ground motion is derived entirely from the realized model's source and attenuation models, the ground-motion fractiles are no longer those of the realized (perhaps uniform-hazard) model if reweighting has occurred.

If model components other than those of some widely-accepted or community or benchmark model are included in a PSHA uncertainty analysis, one needs to determine limits and must determine weights to attach to previously unconsidered or previously rejected or previously averaged models. Informed decisions based on study of the pertinent literature and interaction with experts is required. SSHAC (1997) formalized the process, but their procedures are difficult to implement in practice in no small part because of high costs. OpenSHA (Field et al, 2003) may make it possible to tailor an uncertainty analysis to specific requirements, for example, by incorporating the latest breakthroughs, time-dependency, and so on. Including new components in the uncertainty analysis can significantly alter the resulting distribution of probabilistic ground motions from that which proceeds almost mechanically from a benchmark model. When a uniform hazard PSHA model is available for a region, there is academic and possibly practical benefit in comparing the logic trees that arise from the uniform hazard model with those from alternate models.

This presentation confines itself to the USGS 2002 PSHA model, a model that was updated from 2001 to 2003, described in Frankel *et al.* (2002), and to limited extensions to that model defined above. A desideratum of this report is to maintain the mean hazard of that the USGS model, that is, to only include model extensions that maintain a reasonable agreement of Equation (2) and Equation (1) at all probabilities of interest. For source rates, a necessary constraint on the epistemic weights w_i to apply to individual branches is that $\lambda = \Sigma w_i \lambda_i$. Furthermore, each λ_i is a *mean rate* estimate for a model whose rate is in fact uncertain. If one changes the interpretation to a "1/waiting time to next event" or similar interpretation, the range of possible rates increases dramatically, but no longer conforms to the stationary source-rate assumption of USGS PSHA. An overly high λ_i implies a time-dependent model that promotes the viewpoint that this source is relatively imminent, and an overly low λ_i implies a time-dependent model in which the source is no longer deemed to have much credibility, perhaps because it (or something very much like it, e.g., the Denali, AK mainshock) has recently occurred. While each of these possibilities may be quite plausible and may be promoted by professionally unbiased adherents, they alter the model. Bringing additional information to bear on the seismic hazard estimation problem, such as that which revises mean occurrence estimates, is beneficial and is a central goal of ongoing research, but implies a separate logic-tree analysis from that of the stationary-rate model.

Analogy with Biological Systems

An important feature of a logic tree is that its branches are very different from those of an oak tree. In an oak tree, each branch has an abundance of true information - in fact all of the information needed to generate a copy of the tree. False information, such as genetic mutation, is rare and likely to be deleterious to the health of the tree. In a logic tree, one branch, according to the assumptions of the PSHA, is most accurate, that is, among all the alternative epistemic combinations, only one branch is nearest to the true, underlying model. Other branches typically differ significantly from this most-accurate branch. The logic tree, unlike the oak tree, is burdened with transgenic information. For CEUS sites in the USGS 2002 PSHA, the primary potential errors are associated with attenuation model, magnitude for the characteristic NMSZ or Charleston earthquake, geographic extent of Charleston source zone, location of NMSZ fault, and random seismicity rates at near and regional distances. Geotechnical or site-related variability is frequently of great importance but is outside the scope of this report. A logic tree is mostly a presentation of combinations of poorly to moderately well-constrained and often hotly debated geologic and geophysical model-fitting exercises.

Given the above interpretation, the analysis of the logic-tree branches should include the capacity for identification of the branch components - the DNA of PSHA. We might find some utility in exhibiting the set of model components that combine to yield the median curve and the various fractile curves, such as the 5%, 15%, 85% and 95% curve. For example, if an engineering firm finds difficulty understanding or accepting upper-fractile ground motions, we should like to be able to query the model to determine the combinations of model parameters that produced these motions. Perhaps communication of these alternate models might improve relations between hazard-curve providers and users. Furthermore, knowing the hazard-model components that determine the fractiles might assist in decisions about spending on earthquake-mitigation research.

The Mode of the Distribution of Probabilistic Motions

Some investigators argue that probability-based engineering design should work to an agreed-upon PSHA ground-motion fractile, such as the median or 50% fractile, rather than the mean, used in current standard seismic code specifications. The median is less sensitive than the mean to "outlier" hazard curves, but is more difficult to calculate than the mean (calculated using Equation 1a). Yet another distribution statistic, the mode, might be appropriate in some seismic-design-and-retrofit applications. Although the mode is not conventionally considered, its merits are enumerated below.

Suppose an application requires a level of performance assurance at the most likely probabilistic motion at a given risk level, but that this motion is specified only to a limited precision, such as $x \pm dx$, where x is now the log of the ground-motion variable. If we divide the total range of logged probabilistic motions for the branches into intervals of length $2dx$, then the center of the most probable bin, i.e., the bin with the greatest weight, represents a rational choice - our maximum-likelihood estimate- of x. This most probable bin is the mode of the distribution of probabilistic PGA or SA. The modal bin may or may not cover the median or mean. As is often pointed out, the estimate for the mode depends on the details of the binning, in this case, on the precision dx required. For this report I divide the total range of logarithmic motions into fifteenths, not so much to satisfy any customer need (unknown) but to illustrate how the mode may differ from the median or mean, respectively, and to illustrate how non-central the ground-motion distribution can be for the USGS 2002 PSHA results at some sites at some PE levels. Finer bin widths may separate some significant features that are lumped when using the more common tenths, or deciles.

If there is real utility of the mode of the uncertainty distribution, then the analysis of the branches of the logic tree should be relatively exhaustive or at least should meet some stringent statistical requirements. If only a limited sample of the branches is computed, for example, if a Monte Carlo analysis is performed, it is necessary to demonstrate that the estimated mode is not affected by limited sampling (sampling bias).

If we are permitted to make a few simplifying assumptions in our logic-tree analysis, namely that limited branching on important variables can adequately capture model variability, and, in particular, can adequately model the mode, relatively exhaustive sampling (RES) is a feasible requirement for CEUS sites. For Pacific Northwest sites, after invoking the additional binary fault -system assumption, we find that RES is straightforward. Designing logic trees to achieve RES is much more onerous to implement in parts of California and Nevada where many more combinations of sources - perhaps two to three orders of magnitude more - are implied by the data of the USGS seismic-hazard model. However, in most of the U.S., RES may be a reasonable goal of PSHA logic-tree analysis.

CEUS Sites PSHA Uncertainty

For CEUS sites, some variability in the appearance of the logic tree results from *spatial* variability of significant sources. When an important source is relatively nearby, then branching on that source is increased, while branching on more distant sources is decreased. Details are given in tables below for specific CEUS sites. We note that in the USGS 2002 PSHA, many tectonic features, such as gravity basins and saddles, are not explicitly included in the seismic-hazard model, and are implicitly part of the model only if they are associated with higher rates of background seismicity.

Charleston, South Carolina

For sites in the southeast Atlantic seaboard, Table 1 shows the sources of variability that go into logic-tree calculations of this report, based on the USGS 2002 PSHA. Epistemic weights are given in parentheses.

Table 1. Epistemic variability modeled for Charleston and Clemson sites, and associated weights.

Charleston Mainshock			
Size(M)	Recurrence	Location	Attenuation
6.8(.2)	450 years (.55)	Broad Zone(.5)	AB95 (.25)
7.1(.2)	750 years (.45)	Narrow Zone(.5)	Frankel *et al.* (.25)
7.3(.45)			Toro *et al.* (.25)
7.5(.15)			Somerville *et al.* (.125)
			Campbell (.125)
Background Seismicity			
m->M&Mmax	b-value[*]	a-value	Attenuation
Johnston(1996),7.4 (.5)	0.85(.25)	lowered	AB95 (2/7)
BA87, 7.5 (.5)	0.95(.50)	Frankel *et al.* (2002)	Frankel, *et al.* (2/7)
	1.05(.25)	raised	Toro, *et al.* (2/7)
			Campbell (1/7)
NMSZ Mainshock			
Size(M)	Recurrence	Location	Attenuation
7.3 (.15)	500 years (1)	not varied on branches	AB95 [**]
7.5 (.2)			Frankel *et al.*
7.7 (.5)			Toro *et al.*
8.0 (.15)			Somerville *et al.*
			Campbell

[*]b-value distribution for Charlevoix region differs but is not relevant for South Carolina.

[**]Attenuation model weights for the branch determined once, for Charleston mainshock

Attenuation models above: Atkinson-Boore (1995) (AB95), Frankel *et al.* (1996), Toro *et al.* (1997*)*, Campbell (2003), and Somerville *et al.*, 2001 (*Sea*). Mmax model, BA87=Boore and Atkinson (1987).

Figure 1 shows the 1-s spectral acceleration probability density function for three probabilities: 10%/50 years, left, 2%/50 years, center, and 1%/1000 years, right. The range of the 10%/50 year motion is about 0.02 to about 0.16 g. The range of the 2%/50 year motion is about 0.1 to 0.9 g. The modal bin is from 0.49 to 0.57 g, in this instance above the mean motion for 2% in 50 year PE. The range of the 1%/1000 year motion is about 0.7 to 5.2 g, and the modal bin is from 3.0 to 3.45 g, well above the mean motion, 2.25 g.

Figure 2a shows the median hazard curves and several other fractile curves, 5%, 15%, 85% and 95%, again for the 10%/50 years, solid line, 2%/50 years, dashed, and 1%/1000 years, dotted. Figures 1 and 2a show that the uncertainty range in ground motion is large, with standard deviations near a factor of 2. Figure 2b compares the mean, median and modal hazard for a given 1-Hz spectral acceleration. The mode of the pdf of rate for a given ground acceleration, discretized into fifteenths, has interval probability of 0.15 to 0.2. The mode tracks the mean and median well until about 0.6 g, or $3 \cdot 10^{-4}$ PE, where the modal rate jumps to several times to median rate. The underlying reason for the divergence of these statistics is that the one-corner versus two-corner source models predict very different median ground motions for intermediate to long-period motion. The one-corner models contribute to the modal bin at relatively high ground motions.

We identify the principal epistemic components of select fractile curves in table 2 for the 2% in 50 year PE 1-s spectral acceleration. Note that all the epistemic choices in the model contribute to one or another of the fractile curves, but the wide range is probably dominated by the choice of attenuation function and choice of narrow versus broad zone. The USGS 1-s SA for the 2% in 50 year motion at the Charleston site (32.8°n, 80°w) is 0.393 g. This "map value" is close to the median motion of the uncertainty analysis, i.e., the 50% fractile value. The mean of the 1920 hazard curves for this analysis is 0.405 g. Note that many branches' hazard curves may clump near a given fractile, and that a more complete understanding of sources associated with fractiles would consider a range of solutions near, but not necessarily right at, any given fractile.

Table 2. Principal epistemic components of the hazard curve branch passing through given fractiles for Charleston-site 1-s spectral accelerations having 2% PE in 50 years.

Fractil	Motion	Charleston Mainshock				Gridded Seismicity	
(%)	(g)	M	Zone	Recur.	Atten.	M-cnvrt	b
5	0.141	7.1	broad	750 yrs	AB95	BA87	1.05
15	0.214	6.8	broad	450 yrs	Toro *et al.*	Johnston	1.05
50	0.374	6.8	narrow	450 yrs	Fr *et al.*	BA87	0.95
85	0.583	7.1	narrow	750 yrs	Toro *et al.*	Johnston	0.85
95	0.735	7.3	narrow	450 yrs	Toro *et al.*	Johnston	0.95
100-	0.927	7.5	narrow	450 yrs	Toro *et al.*	BA87	0.85

Whether probabilistic ground motions calculated by the usual algorithm (Equation 1a) are closer in general to the mean or the median of an uncertainty study such as this one is an open question. The answer depends on choices for the branch components in the logic tree and the weights that are applied. The range of probabilistic motions also depends on these details. My experience is that if the analysis confines itself to identified sources of variation in the PSHA, and extensions that are somewhat symmetric around non-varied parameters in the original model, then the median and mean motions are both very close to the usual probabilistic ground motion. Examples will show that beyond their general closeness to one another, there is no simple relationship among these three estimates.

The 5-Hz (0.2 s) SA fractile hazard curves for Charleston are shown in Figure 3. The 10%/50 year curves are solid, the 2%/50 year curves are dashed, and the 1%/1000 year curves are dotted. The median curve for the 2% in 50 year PE has 5-hz motion of 1.605 g, the mean of the 1920 curves is 1.63 g, and the USGS 2002-PSHA value is 1.60 g at this site. In all cases log-log interpolation was performed to obtain the 2% in 50 year motion from sampled motions on the hazard curve. The modal bin for the 2% in 50 year motion (range divided into fifteenths) is 1.54 to 1.69 g. For this PE, the 5-hz mode overlaps the median and mean.

Note that in fig. 3, curves that are at lower fractiles for one probability level are at higher fractiles at other probability levels and vice versa. For instance, the 15% curve for 10%/50 lies below the 5% curve at probability level 10^{-5}. Similarly, the 5% curve for 10%/50 is near the median level at probability level 10^{-5}. Thus, fractile curves need not be sub-parallel over a range of motions.

By identifying the components of a curve we can determine why its rank among the curves changes relatively dramatically with rate. For the 5% curve ordered at a mean rate of 10^{-5}, the SA is 4.58 g, the Charleston-source M is 7.3; its location is in the narrow zone; its recurrence interval is 450 years; the attenuation model is Campbell; and gridded m_{bLg}-to-M is Johnston's. Two main factors explain why this curves dives more rapidly than others shown in fig. 3. First, while the predicted short-period median motion for a given M and distance in the Campbell relation is about the same as that of other attenuation models, the aleatory uncertainty is lower. Therefore, high ground motions occur at more standard deviations above the median, producing a lower rate of exceedances for Campbell versus other attenuation models. Second, the Johnston (J) gridded seismicity component of hazard has a lower Mmax than the Atkinson-Boore (AB) gridded seismicity, corresponding to m_{bLg} 7.0 versus m_{bLg} 7.3 in the eastern seaboard. At very low probabilities, relatively high-M local random seismicity becomes an important contributor to seismic hazard in much of the CEUS. Its greater frequency in the AB model tends to elevate the high-SA part of the hazard curve for branches containing the AB component over those with the J gridded component. The low-SA portion, however, is dominated by relatively low magnitude, but more frequent seismicity. This lower-M seismicity is more dominant in the J model, because the Johnston m_{bLg}-to-M quadratic curve is above that of AB. The phenomenon of logic-tree branch hazard curves crossing one another in CEUS site logic trees is always observed and is primarily due to these dual "Mmax" and m_{bLg}-to-M properties of the input data to the USGS hazard model. The relative size, distance, and return time of characteristic earthquakes on different branches also influences the crossing pattern of logic-tree branch seismic hazard curves.

In the vicinity of the Charleston site there are two alternative seismic sources for large-magnitude events: a concentrated areal source and a broad areal source. The extent to which large ground motions may be likely near the site depends on which source branch is valid and whether one-source-corner or two-source-corner ground motion prediction equations are valid. These distinct features of the PSHA result in alternative shapes of the hazard curves at large ground motions and their tendency to cross, especially for intermediate to long-period spectral acceleration.

Clemson, S.C.

Clemson is a university town in western South Carolina. It is about 300 km from the Charleston source and about 600 km from the NMSZ source. For this logic-tree exercise, Table 1 again describes model variability. A relatively major source of seismic hazard in Clemson is the eastern Tennessee seismic zone (ETSZ), many of whose sources are less than 100 km away. The distribution of seismic hazard estimates at Clemson shows some features of the USGS PSHA that are not well illustrated in the Charleston analysis above.

Figure 4a shows the 1-s SA ground motion uncertainty distributions for the 10%/50 year, the 2%/50 year and the 1%/1000 year PE, respectively. The large overlap of the 10%/50 and the 2%/50 distributions is one significant new feature. Why does this overlap occur? The main reason is that the attenuation models predict very different levels of motion at regional distances - that is, distances of the Charleston and NMSZ sources. The Fea attenuation model predicts that 0.1 g 1-s SA from a NMSZ source is not implausible, at one-to-two σ above the median. The Sea model, on the other hand, predicts that 0.1 g is impossible (more than 3 σ above the median) for the NMSZ source. Similarly the Sea model predicts low likelihood of 0.1 g from a Charleston mainshock source being recorded at a site in Clemson. The AB94 model also predicts low likelihood for this level of motion from both Charleston and the ETSZ, but in their model, this low likelihood is attributed to a two-corner source spectrum rather than to path-attenuation effects. The two Charleston event rates also contribute to the overlap of the 10% in 50 year and 2% in 50 year probabilistic ground motions. Variation of a- and b- values in the truncated GR relation also adds to variability of mean motion. The bimodal shape of the distributions at lower levels of ground motion is accentuated because of large differences in attenuation functions at regional distances.

With respect to probabilistic motion variability, Clemson is representative of many CEUS sites in that it is at regional distance from at least one of the relatively frequent CEUS sources. Thus, we should expect similar overlap in the intermediate- to long-period hazard distributions for the 10% and 2% in 50 year PE motions at many sites in the CEUS. Significant reduction in the variability in estimates of ground motion from regional-distance CEUS sources in the CEUS will require a better understanding of the source process as well as better calibration of low ground motions in regional attenuations. New determination of regional attenuation of Lg from weak-motion recordings from the Advanced National Seismic System in the CEUS may reduce crustal attenuation uncertainty.

Table 3 lists the mean and median hazard for the 2% in 50 year 1 hz and 5-hz SA and PGA, and the USGS 2002 PSHA values for these ground motions. Note the closeness of the mean to the USGS 2002 PSHA. The medians are also close. Note also that the uncertainty ranges are smaller than for Charleston, with standard deviations less than a factor of the square root of 2.

Table 3. Probabilistic motion at Clemson, SC Lat 34.7°n, Long 82.85° w

Frequency:	1 hz SA (g)	5 hz SA(g)	PGA(g)
USGS 2002	0.104	0.350	0.180
Median—this analysis	0.105	0.341	0.171
mode (range)	0.104 to 0.116	0.318 to 0.333	0.161 to 0.168
Mean—this analysis	0.105	0.353	0.179
15% to 85% motion	0.067 to .142	0.304 to 0.427	0.149 to 0.221

Paducah, Kentucky

Table 4 shows the principal sources of ground-motion variability in this analysis for a site in Paducah, a city near the north-eastern edge of the Mississippi embayment.

Background Seismicity			
m->M, Mmax	b-value	a-value	Attenuation (weight)
Johnston(1996),7.4 (.5)	0.95(1)	Fr *et al.* (2002)	AB95 (2/7)
BA87, 7.5 (.5)			Fr *et al.* (2/7)
			Toro *et al.* (2/7)
			Campbell (1/7)
NMSZ Mainshock			
Size(M)	Recurrence	Location	Attenuation
7.3 (.15)	400 years (.58)	western fault (.25)	AB95 (.25)
7.5 (.2)	750 years (.42)	central fault (.5)	Frankel *et al.* (.25)
7.7 (.5)		eastern fault (.25)	Toro *et al.* (.25)
8.0 (.15)			Somerville *et al.* (.125)
			Campbell 2003 (.125)

The NMSZ mainshock is the main source of seismic-hazard variability at all frequencies of ground motion at Paducah, at least for a rock site with no site-specific epistemic uncertainty.

Paducah is about 25 km from the nearest location of the eastern notional NMSZ fault as defined in the USGS 2002 PSHA model. Random seismicity hazard is of limited importance except at very low probability, and Charleston, S.C. hazard is too distant to be of consequence in Paducah.

Source location is not fully determined by knowledge of fault location. For example, a magnitude 7.3 earthquake is not expected to rupture the entire NMSZ fault, whose length in the USGS 2002 PSHA model is about 250 km, although a **M8** earthquake is more likely to do so. NMSZ earthquake scenarios that include partial rupture of the NMSZ fault, while conceptually valid, are not explicitly considered in the USGS 2002 PSHA model, and they are not considered here. However, the 1811-1812 mainshocks, when considered collectively, may have ruptured the entire zone. Previous NMSZ mainshocks may exhibit similar temporal clustering (Tuttle *et al.*, 2002). Implicitly, the USGS 2002 PSHA model may be thought to model some of the ground-motion effects of multiple ruptures, because it uses the nearest distance to the 250 km long fault in source-to-site distance calculations regardless of modeled characteristic magnitude. Several investigations model rupture on specific faults within the NMSZ source zone, many of which are several times as far from Paducah as those considered here. The likelihood of multiple significant shocks over a short time interval (months?) from a restricted source zone provides an additional source of uncertainty for loss studies, one that is not considered in this report.

Figure 5 shows epistemic variability in ground motion for the 1-s SA (left side) and 0.2-s SA (right side) for a NEHRP B-C rock site at Paducah. The distribution of 1-s motion for 2% in 50 year PE overlaps the 10% in 50 year distribution and the 1% in 1000 year distribution significantly. For example, the 2% in 50 year motion can be as low as 0.1 *g*, and the 10% in 50 year motion can be as high as 0.25 *g*. Much of the overlap at intermediate to long-period ground motion is due to uncertainty about the source spectrum: 1-corner or 2-corner. On the other hand, the 0.2-s motion probabilistic motion distributions at these three probabilities of exceedance have very limited overlap. The principal sources of variability are source location, attenuation model and NMSZ source recurrence time. These are more fully explored in Cramer (2001) and Newman *et al.*(2001). Table 5 lists the mapped values, the median, the modal bin range, and the mean motion for Paducah and three ground motion frequencies. At Paducah, the modal-bin motion is higher than the mean or median motion because of the closeness of several attenuation model predictions of near-source motion from the NMSZ mainshock.

Table 5. Probabilistic motion at Paducah, KY. Lat 37.1 °n, Long 88.6° w

Frequency:	1 hz SA (g)	5 hz SA	PGA
USGS 2002	0.467	1.70	0.918
median-this analysis	0.443	1.70	0.913
mode (range)	0.49 to 0.57	1.90 to 2.09	1.05 to 1.15
mean - this analysis	0.468	1.72	0.919
15% to 85% range	0.30 to 0.65	1.21 to 2.13	0.70 to 1.16

In a more thorough evaluation of seismic hazard uncertainty at Paducah, the effects of the sediment column and the effects of the Mississippi embayment need to be accounted for. Some soil-column effect on the probabilistic motion at Paducah is discussed in Cramer (2003). Because Paducah is near the edge of the Embayment, basin-margin-generated surface waves may be important to intermediate- to long-period seismic hazard at Paducah. Such site-specific details of uncertainty are beyond the scope of the USGS 2002 PSHA model and this report.

Pacific Northwest Sites PSHA Uncertainty

Many recent earth-science investigations have resulted in significant enhancements of PSHA models for the Pacific Northwest, including GPS work, studies of the subduction process, crustal structure and deformation studies. Revisions documented in the USGS 2002 PSHA report (Frankel *et al.*, 2002) were made to several categories of sources in the Pacific Northwest, including random shallow seismicity, random deep seismicity, seismicity on mapped faults, and hazard from Cascadia megathrust and other Cascadia sources. Several attenuation models were added to those considered in the USGS 1996 PSHA report (Frankel *et al.*, 1996). Effects on seismic hazard from many of these revised models are studied in Petersen *et al.* (2002). Sources of variability in Petersen *et al.* (2002) overlap significantly, but not completely, those in the USGS 2002 hazard maps.

For this report, we focus on the models that actually go into the preparation of the USGS 2002 national seismic hazard maps. Extensions defined here are (1) faster and slower rates of subduction compared to those of the 2002 report, (2) faster and slower fault activity rates, and lower and higher b-values for (3) random shallow and (4) deep seismicity. Table 6 shows the sources of variability for seismic hazard in the Pacific NW that are considered in this report. In table 6, T is the mean interval for characteristic earthquakes on the specified fault, as given in the Quaternary fault database, found on the web at http://geohazards.cr.usgs.gov/qfaults/index.html. To model uncertainty in T, the mean rate is varied to yield a factor of two in the range, $0.8\,T$ to $1.6\,T$.

Table 6. Principal sources of uncertainty in Pacific Northwest seismic hazard model.

Cascadia Subduction			
Size(M) (weight)	Recurrence (weight)	Location (weight)	Attenuation(weight)
8.3(0.5)	400 years **M9** (.6)	Modified '96 (.5)	Sadigh *et al.* (0.5→0)[**]
9.0(0.5)	800 years **M9** (.4)	Top transition (.1)	Youngs *et al.* (.5→1)
		Middle transition (.2)	
	similar for **M8.3**	Bottom transition (.2)	
Gridded Shallow Seismicity			
Zones (weight)	b (weight) , a-value	Location	Attenuation

Puget/GPS (.5)	0.85(.25), lowered	strike randomized	A&S (0.25)
WUS/Catalog (.5)	0.95(.50), same	for **M**≥6	Sadigh *et al.* (0.25)
	1.05(.25), raised	Depth 5 km (1)	BJF (0.25)
			C&B (0.25)

Gridded Deep Seismicity

Zones	B (weight), a-value	Fixed Parameters	Attenuation
PacNW (1)	0.3 (0.25), lowered	Mmax 7.2 (1)	A&B, global (0.25)
N. Calif. (1)	0.4 (0.5), same	Depth 50 km (1)	A&B, Cascadia (0.25)
	0.5 (0.25), raised		Youngs *et al.* (0.5)

Main Characteristic Fault(s) (varies with site)

Size(M) (weight)	Recurrence(weight)	Fixed Parameters	Attenuation(weight)
M from W&C[*] (.6)	1.6 T (0.4)	Location	A&S (0.25)
M–0.2 (.2)	0.8 T (0.6)	Dip	Sadigh *et al.* (0.25)
M+0.2 (.2)		Depth	BJF (0.25)
			C&B (0.25)

Main Truncated GR Fault(s)

Size(M)	Recurrence	Fixed Parameters	Attenuation
6.5 to **M**–0.2(0.6)	1.6 T (0.4)	Location	A&S (0.25)
6.3 to **M**–0.4 (0.2)	0.8 T (0.6)	Dip	Sadighea (0.25)
6.7 to **M** (0.2)		Depth	BJF (0.25)
			C&B (0.25)

Other Characteristic Fault(s)

Size(M)	Recurrence	Fixed Parameters	Attenuation
M from W&C (0.6)	1.6 T (0.4)	Location	A&S (0.25)
M–0.2 (0.2)	0.8 T (0.6)	Dip	Sadigh *et al.* (0.25)
M+0.2 (0.2)		Depth	BJF (0.25)
			C&B (0.25)

Other Truncated GR Fault(s)			
Size(M)	**Recurrence**	**Fixed Parameters**	**Attenuation**
6.5 to **M**–0.2 (0.6)	1.6 T (0.4)	Location	A&S (0.25)
6.3 to **M**–0.4 (0.2)	0.8 T (0.6)	Dip	Sadigh *et al.* (0.25)
6.7 to **M** (0.2)		Depth	BJF (0.25)
			C&B (0.25)

*W&C=Wells and Coppersmith (1994). A&B= Atkinson and Boore (2003), A&S=Abrahamson and Silva (1997), BJF= Boore *et al.* (1997), C&B=Campbell and Bozorgnia (2003), Also, Sadigh *et al.* (1997), and Youngs *et al.* (1997) are referenced above.

** For Cascadia sources, attenuation model weight is a function of distance, discussed below.

The Pacific NW seismic-hazard logic tree of this report has over 92,000 branches. Sources of variability are well sampled, but not exhaustively. Two guiding principles are (1) to include potential extreme combinations (e.g., fast or slow rates for both fault-groups on a branch) and central models (e.g., central magnitudes for both fault-groups on a branch), but to omit some intermediate models, and (2) to exhaustively sample Cascadia variability relative to other sources.

In spite of the greater number of sources and branches in the Pacific NW analysis compared to the CEUS, the computed range in uncertainty, for example, the difference between the 5% and 95% fractile curves, is not greater in the Pacific Northwest than in the CEUS. This is because there are many more strong-ground-motion records and consequently less variability in the ground-motion attenuation models for the WUS than for the CEUS, at least for crustal sources. Also, the USGS 2002 PSHA does not explicitly include a two-corner source model, such as Atkinson and Silva (2000), among its shallow-source attenuation models for sites in the WUS, although the empirical models that are used track the Atkinson and Silva (2000) medians at intermediate to long-period SA. In the CEUS, the two-corner source model is a major contributor to PSHA variability at least for long-period (>2s), and to a lesser extent, intermediate-period (≈1s) SA. Furthermore, there is not much uncertainty in magnitude for characteristic sources on faults or in M_{max} for deep seismicity in the Pacific NW model. However, as pointed out in Petersen *et al.* (2002), larger magnitude intraplate events have been recorded in Mexico (**M**7.8) and other subduction zones than are assumed possible in USGS 2002 PSHA Pacific NW and northern California models.

In Table 6, effects from California sources have been omitted, as well as those from the Basin and Range province. We therefore confine analysis to sites in Washington and north-western to north-central Oregon.

Seattle

Figure 6 shows three 1-s SA distributions for a site in Seattle for three PE s. The central distribution is for the 2% in 50 year PE. This distribution is reasonably symmetric and the mean motion of the curves matches the USGS 2002 PSHA map motion, 0.499 g, very well. The 15% and 85% motions for this Seattle site at the 2% in 50 year PE are 0.434 and 0.565 g, respectively. The shape of the uncertainty distribution is much more asymmetric at 1% in 1000 year PE. The U.S.G.S. did not publish a map value for 1% in 1000 year PE, so we do not compare map motion with mean motion. Some variations in fractile curves should be expected for different sites in Seattle, primarily as a function of distance to the Seattle fault.

Figure 7 shows the 0.2-s SA motion distributions for the Seattle site. Again the 2% in 50 year distribution is reasonably symmetric and the mean value is very close to the map value, 1.477 g. The 15% and 85% fractile motions are 1.20 and 1.79 g, respectively. An as-yet unquantified degree of symmetry in the ground-motion uncertainty distribution seems necessary to insure the approximate equality of "mean" motions as defined by equations (1) and (2).

Spokane

Figure 8 shows the PSHA ground-motion uncertainty pdfs for a site in Spokane for 1-s SA. As in previous figures, three PEs are considered, 10% in 50 years, 2% in 50 years, and 1% in 1000 years. Note that the 1-s SA for the 2% in 50 year PE has a low-ground-motion lobe. The USGS 2002 map motion, 0.113 g, is slightly higher than the median, and is in good agreement with the Equation 2 mean. The absence of a bias of the form (Equation 1 mean motion) < (Equation 2 mean motion) for this asymmetric distribution illustrates an important feature of Equation 1 and of conventional PSHA estimates that utilize it. Equation (1) tends to agree with Equation 2 estimates of "the" mean when the ground-motion uncertainty distribution is symmetric or when there is a lobe of probability at a low ground-motion. The "Info" quantity in figure 8 is information density, a measure of the distance of a distribution from the uninformative uniform distribution, which has 0 "Info."

Figure 9 shows ground-motion uncertainty pdfs for 0.2-s SA. For low PE, the 0.2-s SA distribution is almost uniform. Without gridded b-value variability, the Spokane 5-hz density function is 4 spikes, one corresponding to each of the attenuation models used for crustal sources in the Pacific Northwest. While including b-value or other forms of gridded-hazard variability smoothes the distribution somewhat, there is a prominent lobe in fig. 9. The distribution is reasonably symmetric, nevertheless, and the agreement between mapped motion, median motion and mean motion is outstanding at Spokane for the 0.2-s SA. The reason that the 1-s distribution is more complex than the 0.2-s distribution in Spokane is that distant Cascadia subduction sources may contribute significantly to intermediate to long-period hazard. Cascadia sources are of some importance to the 0.2-s motions at high probability of exceedance (10% in 50 years, for example) but Cascadia's influence tapers off quickly at higher ground motions. The simplicity of the 5-hz distribution results from the lack of relatively close fault sources, limited gridded-source variability (Puget lowlands is too far), and the attenuation of Cascadia sources at higher frequencies. For Spokane, the only earthquakes that are capable of contributing > 0.3 g are local gridded sources, and most of the variation in 0.2-s ground motion is due to attenuation models. At very low PE, local gridded sources are also the only contributors to 1-s SA in the USGS 2002 PSHA model.

Because we use the same attenuation model for all crustal sources on a logic-tree branch, we can report the conditional mean motion for a given PE and for a given attenuation model. For the Spokane site, for 2% in 50 year PE, and for 5-hz SA, this conditional mean is 0.33 *g* using Boore *et al.* (1997), 0.40 *g* using Campbell and Bozorgnia (2003), 0.43 *g* using Sadigh *et al.* (1997), and is 0.46 *g* using Abrahamson and Silva (1997). These conditional mean motions are computed from equation 2, constrained to use one attenuation model for all continental sources.

Central and non-Central Tendencies

For sites in the Pacific Northwest, the mapped probabilistic motions approximately match the mean motion from the logic-tree branches. Less than 2% difference is found at most Pacific Northwest cities, for a broad range of spectral periods and PEs. The medians of the logic-tree branches at a fixed PE also are in agreement with the mapped motions, generally. However, the coastal region of Washington and Oregon shows some significantly lower mean ground motions than those of the PSHA map. We did not find this mismatch for CEUS sites that we studied. We always found good agreement of the Equation (1) and Equation (2) probabilistic motions for CEUS sites. These anomalous results in parts of the Pacific NW are not difficult to understand.

The mismatch-of-motion phenomenon at coastal cities is worth examining because PSHA motions are said to be mean motions. Table 7 shows the USGS 2002 PSHA map values, available on the Web at http://eqhazmaps.usgs.gov/html/data2002.html, the median branch value, two mean estimates, the modal-bin range for 1-s and 0.2-s SA and for PGA for two coastal cities, Tillamook, Oregon and Aberdeen, Washington. Corresponding data for sites in several inland cities, including Portland and Seattle, are also shown in Table 7. The mode and the first mean use horizontal averaging (computed at fixed probability). The conventional mean (computed at a fixed ground motion) is given in parentheses with a "V." For binning, the total range of probabilistic motion variability is divided into fifteenths, as it was in the CEUS.

Table 7. USGS 2002 PSHA (Map) motions, 2% PE in 50 years, and logic-tree central-tendency statistics, Pacific NW sites.

Tillamook, OR(1)	1s SA (*g*)	0.2s SA	PGA
Map	0.66	1.31	0.55
Median	0.55	1.23	0.51
Mean	0.61 (0.67V)	1.22 (1.33V)	0.50 (0.55V)
Modal Bin	0.50-0.55	1.71-1.88	0.69-0.76
Aberdeen, WA (2)	1s SA (*g*)	0.2s SA	PGA
Map	0.71	1.43	0.60

Median	0.57	1.36	0.58
Mean	0.66 (0.71V)	1.37 (1.43V)	0.58 (0.60V)
Modal Bin	0.54-0.58	1.36-1.46	0.55-0.60
Portland, OR (3)	1s SA (*g*)	0.2s SA	PGA
Map	0.34	0.99	0.42
Median	0.33	0.94	0.40
Mean	0.33 (0.33V)	0.95 (0.95V)	0.40 (0.40V)
Modal Bin	0.35-0.37	0.95-1.01	0.41-0.43
Salem, OR(4)	1s SA (*g*)	0.2s SA	PGA
Map	0.34	0.80	0.33
Median	0.32	0.78	0.32
Mean	0.34 (0.34V)	0.79 (0.80V)	0.33 (0.33V)
Modal Bin	0.30-0.32	0.71-0.76	0.31-0.32
Seattle, WA (5)	1s SA (*g*)	0.2s SA	PGA
Map	0.50	1.48	0.65
Median	0.49	1.41	0.63
Mean	0.50 (.50V)	1.47 (1.49V)	0.64 (0.66V)
Modal Bin	0.49-0.52	1.30-1.41	0.60-0.65
Spokane, WA (6)	1s SA (*g*)	0.2s SA	PGA
Map	0.11	0.40	0.17
Median	0.10	0.40	0.16
Mean	0.11 (0.11V)	0.40 (0.40V)	0.17 (0.17V)
Modal Bin	0.09-0.10	0.39-0.41	0 .19-0.20
Everett, WA (7)	1s SA (*g*)	0.2s SA	PGA
Map	0.40	1.15	0.51
Median	0.40	1.14	0.51
Mean	0.40 (0.40V)	1.14 (1.15V)	0.50 (0.51V)

Modal Bin	0.41-0.43	1.11-1.17	0.50-0.53
Port Angeles (8)	1s SA (*g*)	0.2s SA	PGA
Map	0.48	1.12	0.50
Median	0.45	1.12	0.51
Mean	0.46 (0.48V)	1.12 (1.13V)	0.50 (0.50V)
Modal Bin	0.44-0.47	1.15-1.23	0.50-0.54

(1) 123.85° W 45.45° N (2) 123.8° W 47.0° N (3) 122.65° W 45.5° N (4) 123.05° W 44.95° N
(5) 122.35° W 47.6° N (6) 117.45° W 47.65° N (7) 122.2° W 48° N (8) 123.45° W 48.1° N

To understand the mismatch between mapped motion and median or mean motion at a fixed PE at coastal sites in the Pacific NW, we compare the 1-s probabilistic-motion distributions for the 2% in 50 year PE for Tillamook, OR and Port Angeles, WA, in Figure 10a. Port Angeles is a port city on Puget Sound about 50 miles from the coast. In figure 10a, the Port Angeles distribution is more concentrated than that for Tillamook, with a small side-lobe at 0.7 *g*. The Tillamook distribution is spread out, with a sizable, almost modal, side-lobe at 1.0 *g*. Again, the "Info" quantity in figure 10a is information density, a measure of the distance of a distribution from the uninformative uniform distribution, which has 0 "Info." The suggestion of this figure is that source/attenuation elements that produce relatively extreme motion are effectively weighed more heavily in Equation (1) than in Equation (2). Figure 10b again examines the 1-s seismic hazard at Tillamook. In this figure, the mean, median and modal hazard curves are compared. The mean is computed from Equation (1b), i.e., the vertical average of rates for a fixed spectral acceleration. The mode is computed from the 1/15 of the vertical range that has the greatest epistemic weight. The fact that the mode does not mimic the mean, but jumps from above the mean to significantly below indicates that the Tillamook PSHA models have a bi-modal or multi-modal distribution. Figure 10c examines the 1-Hz seismic hazard at Port Angeles again. The estimates for mean, median and mode are computed at fixed SA values following conventional methodology. This plot is very typical of site hazard curves studied for this report. That is, the mean, median, and mode tend to track one another quite closely unless highly divergent epistemic models with significant weight populate the logic-tree branches.

Figure 11 shows the distributions of uncertainty in PGA for the same pair of sites. Again Port Angeles exhibits a more concentrated distribution of possible PGA values than does Tillamook, this time without an upper 1/15 ground-motion lobe. In this instance Port Angeles mean PGA is only 1% lower than its mapped PGA. Tillamook's PGA distribution has a sizable lobe at the extreme value of 0.7 *g*, and Tillamook's mapped PGA (equation 1a) is over 10% greater than its mean PGA (equation 2). At Pacific coast sites, then, the USGS 2002 PSHA model has components that may produce a significant concentration of hazard at the upper 7% tail, perhaps without corresponding components in the lower tail. At sites where this occurs, the motion computed using Equation (1) may be 10% to 15% greater than that using Equation (2). In most

PSHA studies, vertical averages (equation 1b) are more strongly influenced by the upper fractiles than horizontal averages (equation 2), because of the increased range in numbers vertically. In the vertical averages, a lower-tail fractile is irrelevant, since it cannot compensate. When computing the horizontal average, lower-fractile components may bring down the average significantly.

The Sadigh *et al.* (1997) ground motion relationship when applied to predict motion from Cascadia subduction earthquakes is the underlying reason for this lobe. Seismic hazard resulting from Cascadia subduction at sites on the western coast of North America is modeled using the Geomatrix attenuation model (Youngs *et al.*, 1997) for subduction sources (50% weight) and the Sadigh *et al.* relation (50% weight). The Sadigh *et al.* relation was designed for crustal events but is used as well for subduction events when source-to-site distance is relatively short. Its weight in the USGS 2002 PSHA model tapers to 0 at distance 70 km. Pacific NW coastal site probabilistic motions tend to be relatively high because two of the four models of Cascadia subduction are close enough to coastal sites for Sadigh *et al.* to receive maximum weight, and the other two Cascadia edge locations (top of transition zone and middle, respectively) are close enough to receive some weight. Atkinson and Boore (2003) modeled a vastly expanded set of strong ground motion records for subduction events, compared to Youngs *et al.* (1997). Their regression model confirms the Youngs *et al.* conclusion that near-source median ground motions from subduction sources are significantly less than those from crustal events of a comparable magnitude, although the updated data-base remains deficient in Cascadia-specific subduction sources. The rationale for using the Sadigh *et al.* relation with all Cascadia subduction sources in the USGS 2002 PSHA is discussed on p. 16 of Frankel *et al.* (2002), and is not repeated here.

We may compare average ground motions conditional on Cascadia attenuation model. For Tillamook, for the 5-Hz SA, and for 2% /50 year PE, the mean SA given Sadigh *et al.* (1997) is 1.38 *g*, but is only 1.10 *g* given Youngs *et al.* (1997). For the 1-Hz SA and the same PE, the contrast in mean motions conditional on these two relations is even greater: 0.79 *g* SA given Sadigh *et al.* (1997), and 0.46 *g* given Youngs *et al.* (1997). We may also compare ground motions conditional on Cascadia source location. See figure 7 of Frankel *et al.* (2002). For the model in which seismic rupture stops at the top of the transition zone, the 2%/50 year PE 1-s mean SA is 0.34 *g*, whereas for the model in which seismic rupture stops at longitude -123.8°, the 1-s mean SA is 0.73 *g*. Thus, future progress to reduce uncertainty in attenuation model median motion and extent of downdip seismogenic rupture of the subducting slab both may be expected to strongly affect estimates of mean and median probabilistic motion at coastal Pacific NW sites.

Portland Oregon entries in Table 7 also indicate slightly higher high-frequency map probabilistic motion, compared to Equation 1b and Equation 2 mean of the logic-tree branches. The probabilistic motion distributions for Portland for the 2% in 50 year PE are slightly asymmetric, skewed but without the prominent lobe at the upper 1/15 ground motions that are seen in the Tillamook distributions. The Sadigh *et al.* relation/ Cascadia source combination does not contribute to variability of ground motion at Seattle or Portland. Both cities are more than 100 km from the nearest edge of any of the models of the subducting slab. Discrepencies between Equation 1a and 1b mean estimates are in part due to ground-motion sampling differences. In this paper, density of ground-motion samples was purposefully increased compared to the USGS 2002 PSHA to reduce potential hazard-curve interpolation error.

Summary and Conclusions

Seismic hazard model variability is analyzed using logic-tree methodology for sites in the CEUS and the Pacific NW, using the components of epistemic variability that can be identified in the USGS 2002 PSHA input model. An additional source of uncertainty not explicitly included in the USGS 2002 PSHA model is mean rate of occurrence variability for faults and Cascadia subduction. To represent this variability, mean rates are bifurcated: $r_1=1.2r$ and $r_2=0.6r$. Furthermore, gridded-source truncated GR earthquake rates are redistributed over magnitudes of engineering interest by varying b-value. For these sources, the total expected number of events N in each grid-cell is fixed at the USGS 2002 PSHA model value.

At many CEUS sites, logic-tree seismic hazard curves show a greater range than those for Pacific NW sites, at least for motions at the 2% in 50 year PE. Much of the higher variability in the CEUS is due to the inclusion of one-corner and two-corner source models in the CEUS attenuation models, and much is due to a considerable range of expert opinion about the magnitude of NMSZ and Charleston, SC mainshocks. At CEUS sites, the USGS 2002 PSHA input model explicitly contains very little information on epistemic uncertainty of earthquake rates, although it must be emphasized that the purpose of that model is to compute mean rates of exceedance, not fractile rates. In this article, probabilistic motion variability at CEUS sites at some distance from these identified characteristic source regions is generated by varying b-value while fixing N, but other approaches should also be considered. For example, the USGS determines a-grids (i.e., rates of magnitude $0\pm dm$ earthquakes) based on earthquake catalogs with different m_b minima (3.0, 4.0, and 5.0). The a-grids associated with minimum $m_b =5.0$ have broad regions with $a=0$, because there are relatively few $m_b{\geq}5$ earthquakes in the CEUS catalog. Hazard curves from branches that use such a-grids can yield probabilistic ground motions that are 10 times less than the mean. However, there may be sound geophysical reasons for omitting such models from the logic tree, such as the long expected return time of M5+ earthquakes compared to the length of the CEUS historical record.

Hazard curves associated with logic-tree branches may have considerably different slopes as a function of probability of exceedance. This implies that the seismic-hazard model that happens to be at the median of the distribution at one PE is not necessarily at or near the median at another PE. This paper identifies all of the important hazard components that go into the median-hazard model (and other fractile models) because we believe that this information may be helpful to users of PSHA data. However, users should keep in mind that these component models are likely to change with rising PE.

In the Pacific Northwest, variability in probabilistic motion is often strongly a function of site location relative to Cascadia subduction events. At Tillamook and Aberdeen, details about how ground motion from Cascadia sources is modeled produce almost all of the extremal probabilistic motions. In Spokane, WA, Cascadia is too distant to contribute much to uncertainty, and the principal hazard variability is due to gridded or background seismicity rate uncertainty and to attenuation model uncertainty for relatively close sources. Source uncertainty is partitioned into overall rate of potentially damaging earthquakes, relative rates of larger to smaller earthquakes within that set, and the maximum magnitude of possible earthquakes within a limited radius of the site. Diverse estimates may be found for each of these parameters.

This report shows the range of probabilistic motions that are possible by combining in sensible ways the models of epistemic uncertainty that are in the USGS 2002 PSHA model. However, at many conterminous U.S.A. sites, the expected variability in ground motion is due to uncertainty in the long-term rates of random seismicity at magnitudes in the 5 to 7 range. A more thorough treatment of this factor requires additional catalog re-sampling and other statistical analysis that is beyond the scope of this report. Inclusion of such modeled uncertainty is likely to extend the low ground motion parts of the distribution, for reasons previously discussed. This will decrease the relative compactness and symmetry of the reported distributions. This will also increase the discrepancy between means according to equation (1) and equation (2). However it can be argued that basing epistemic uncertainty on resampled individual *parts* of the historical catalog may exaggerate the uncertainty, if most hazard input models average over the spatial distribution of such parts.

The universe of PSHA models is divided into conceptual models and realized models. A realized model will probably have peer review and acceptance by customers, and, in some cases its data may be adopted into building codes as the basis for a spatially uniform level of risk. Conceptual models may be bounded by the limits of informed imagination, or may be restricted to expert models used in special site studies. Much controversy, difficulty and expense in logic-tree analysis is associated with how much conceptual to intersperse with realized model components. A forward-looking desideratum is that all PSHA input files explicitly contain summary information about model variability for all significant factors. For example, it is sufficient to average mean source rate-of-occurrence estimates if the only goal is to estimate mean exceedance rates, but variability of expert opinion of rate estimates may be of fundamental importance in assessments of variability of mean seismic hazard. A good example is variability in the mean recurrence time of NMSZ main shocks. The mean rate is given as 500 years in the input files for the USGS 2002 PSHA, but variability in the mean is found to be important in Cramer (2001) and Newman *et al.* (2001). A reasonable goal is to be able to construct - if not automatically at least without gargantuan effort - an unbiased epistemic uncertainty analysis from the input files and computer codes that go into the production of a mean seismic hazard analysis, for example, the next national seismic hazard map update. The ability to automatically perform logic-tree analysis is also alledged to be a goal of Open SHA (Field *et al.,* 2003).

Sometimes it is suggested that seismic hazard mitigation should be based on a given logic-tree branch fractile rather than on the mean motion for a given PE. The 50% fractile appears remarkably close to the mean hazard for the work presented here, and therefore provides little additional information. This article suggests that the mode of the uncertainty distribution of binned ground motions may be a useful quantity to characterize the probabilistic hazard for a given PE, although in this study, the modal estimates are for the most part sufficiently close to the means that they tend to confirm the usefulness of those means as target motions for seismic-resistant design.

Acknowledgments

Mark Petersen, USGS, Ken Campbell, ABS Consulting, and Craig Taylor, co-editor, provided many valuable suggestions to improve drafts of this document. Further technical reviews by Art Frankel and David Perkins, USGS, are appreciated. Errors and conclusions are the sole responsibility of the author. Implied points of view may not be shared by USGS management.

References

Algermisson, S. T., D. Perkins, P. Thenhaus, S. Hanson, and B. Bender (1990). Probabilistic earthquake accceleration and velocity maps for the United States and Puerto Rico, U.S. Geological Survey, Miscell. Field Studies Map MF-2120.

Atkinson, G. and D. Boore (1995). Ground motion relations for eastern North America, *Bull. Seism. Soc. Am.*, **v** 85, pp. 17-30.

Atkinson, G. and D. Boore (2003). Empirical ground-motion relations for subduction-zone earthquakes and their application to Cascadia and other regions, *Bull. Seism. Soc. Am.*, **v 93,** 1703-1723.

Atkinson, G. and W. Silva (2000). Stochastic modeling of California ground motions, *Bull. Seism. Soc. Am.*, **v** 90, pp.255-274.

Bommer, J., N. Abrahamson, F. Strasser, A. Pecker, P. Bard, H. Bungum, F. Cotton, D. Fäh, F. Sabetta, F. Scherbaum, and J. Studer (2004). The challenge of defining upper bounds on earthquake ground motions, *Seism. Res. Lett.*, **v** 75, 82-95.

Boore, D.M., and G.M. Atkinson (1987). Stochastic prediction of ground motion and spectral response parameters at hard-rock sites in eastern North America, *BSSA*, v.77, 440-467.

Brune, J. N. (2000). Precarious rock evidence for low ground shaking on the footwall of major normal faults, *Bull. Seism. Soc. Am.*, **v** 90, pp. 1107-1112.

Campbell, K. W. (2003) Prediction of strong ground motion using the hybrid empirical method and its use in the development of ground-motion (attenuation) relations in eastern North America. *Bull. Seism. Soc. Am.*, **v** 93, pp 1012-1033.

Campbell, K.W., and Y. Bozorgnia, 2003. Updated near-source ground motion (attenuation) relations for the horizontal and vertical components of peak ground acceleration and acceleration response spectra, . *Bull. Seism. Soc. Am.*, **93**, 314-331.

Cornell, C. Allin (1993). Which 84th percentile do you mean? Fourth DOE Natural Phenomena Hazards Mitigation Conference - 1993. Department of Energy report.

Cramer, C. H. (2001). The New Madrid Seismic Zone: Capturing variability in seismic-hazard analyses, *Seism. Res. Lett.* **v** 72, pp 664-672.

Cramer, C. H. (2003). Site-specific seismic-hazard analysis that is completely probabilistic, *Bull. Seism. Soc. Am.*, **v** 93, 1841-1846.

Dolan, J., D. Bowman, and C. Sammis (2003). Paleoseismic evidence for long-term and long-range elastic interactions in southern California, *Proceedings and Abstracts,* **V.** XIII, Southern California Earthquake Center, abs. p 49.

Electric Power Research Institute (EPRI), 1988. Seismic hazard methodology for the central and eastern United States, v 1, part 1: Theory. Prepared by Jack R. Benjamin and Associates, Risk Engineering, Inc., Woodward-Clyde Consultants, and C.A. Cornell. Palo Alto, California.

Field, N., T. Jordan, and C.A. Cornell (2003). OpenSHA: A developing community-modeling environment for seismic hazard analysis. *Seism Res. Lett.* **v** 74, 406-419.

Flueck, P., R. D. Hyndman, and K. Wang, 1997. Three-dimensional dislocation model for great earthquakes of the Cascadia subduction Zone, *J. Geophys. Res.*, v. 102. pp 20539-20550.

Frankel, A., C. Mueller, T. Barnhard, D. Perkins, E. Leyendecker, N. Dickman, S. Hanson, and M. Hopper (1996), National seismic-hazard maps: documentation, June 1996, U.S. Geol. Surv. Open-File Rept., 96-532, 110 pp.

Frankel, A., Harmsen, S., Mueller, C., Barnhard, T. , Leyendecker, E., Perkins, D., Hanson, S., Dickman, N., and Hopper, M, 1997. USGS National Seismic Hazard Maps: uniform hazard spectra, de-aggregation, and uncertainty, in *Proceedings of the FHWA/NCEER Workshop on National Representation of Seismic Ground Motions for New and Existing Bridges, NCEER publication* 97-0010.

Frankel, A. D., M. D. Petersen, C S. Mueller, K. M. Haller, R. L. Wheeler, E.V. Leyendecker, R. L. Wesson, S. C. Harmsen, C. H. Cramer, D. M. Perkins, and K. S. Rukstales (2002). Documentation for the 2002 Update of the National Seismic Hazard Maps, U.S. Geological Survey Open-File Report 02-420.

Harmsen, S., D. Perkins, and A. Frankel, 1999. Deaggregation of Probabilistic Ground Motions in the Central and Eastern United States, *Bull. Seism. Soc. Am.* **89**, 1-13.

Harmsen, S., 2001. Mean and modal epsilon in the deaggregation of probabilistic ground motion, *Bull. Seism. Soc. Am.* **91**, 1537-1552.

Johnston, A.C., K.J. Coppersmith, L.R. Kanter, and C.A. Cornell, 1994. The earthquakes of stable continental regions: assessment of large earthquake potential, EPRI TR-102261, J.F. Schneider, ed., Electric Power Research Institute, 309 pp.

Johnston, A.C., 1996. Seismic moment assessment of earthquakes in stable continental regions-I. Instrumental seismicity, *Geophys. J. Int.* **124**, 381-414.

King, G.C.P., 1983. The accommodation of strain in the upper lithosphere of the earth by self-similar fault systems: the geometrical origin of *b*-value, *Pure Appl. Geophys.*, **121**, pp. 761-815.

Leyendecker, E.V., J. Hunt, A. Frankel, and K. Rukstales, 2000. Development of Maximum Considered Earthquake Ground Motions, *Earthquake Spectra* **16**, pp21-40.

Mueller, C. S., A. D. Frankel, M. D. Petersen, and E. V. Leyendecker, 2003. Probabilistic Seismic hazard maps for Puerto Rico and the U.S. Virgin Islands, *Seism. Res. Lett.* **74**, p. 207.

Newman, A., J. Schneider, S. Stein, and A. Mendez (2001). Uncertainties in seismic hazard maps for the New Madrid Seismic Zone and implications for seismic hazard communication, *Seism. Res. Lett.* **v** 72, pp. 647-663.

Petersen, M. D., C. H. Cramer, and A. D. Frankel, 2002. Simulations of seismic hazard for the Pacific Northwest of the United States from earthquakes associated with the Cascadia subduction zone, *Pure appl. geophs.*, **v** 159, pp. 2147-2168.

SSHAC, Senior Seismic Hazard Analysis Committee (1997). Recommendations for probabilistic seismic hazard analysis: guidance on uncertainty and use of experts, U.S. Nuclear Regulatory Commission, U.S. Dept. of Energy, Electric Power Research Institute; NUREG/CR-6372, UCRL-ID-122160, Vol 1-2.

Sadigh, K., C.-Y. Chang, J. A. Egan, F. Makdisi, and R.R. Youngs, 1997. Attenuation relationships for shallow crustal earthquakes based on California strong-motion data, *Seism. Res. Lett.*, 68, 180-189.

Somerville, P. N., N. Collins, N. Abrahamson, R. Graves, and C. Saikia (2001). Ground motion attenuation relations for the central and eastern United States, final report to U. S. Geological Survey.

Thenhaus, P. and K. Campbell, 2002. Seismic hazard analysis, in *Earthquake Engineering Handbook* (W.F. Chen and C. Scawthorn, eds.), CRC Press, Boca Raton, Florida. Chapt. 8, 1-50.

Toro, G., Abrahamson, N., and J. Schneider, 1997. Model of strong ground motions from earthquakes in central and eastern North America: Best estimates and uncertainties. *Seism. Res. Lett.*, **68**, 41-57.\

Tuttle, M., E. Schweig, J. Sims, R. Lafferty, L. Wolf and M. Haynes, 2002. The earthquake potential of the New Madrid Seismic Zone, *Bull. Seism. Soc. Am.*, **v** 92, 2080-2089.

Wells, D. L, and K. J. Coppersmith, 1994. New empirical relationships among magnitude, rupture length, rupture width, rupture area and surface displacement, *Bull. Seism. Soc. Am.*, **84**, 974-1002.

Wesson, R.L., A.D. Frankel, C. S. Mueller, and S.C. Harmsen, 1999. Probabilistic seismic hazard maps of Alaska, U.S. Geol. Surv. Open-File Report 99-36, 43 p.

Working Group on California Earthquake Probabilities (2002). Earthquake probabilities in the San Francisco Bay region: 2002-2031 – *U.S. Geological Survey Circular* 1189.

Youngs, R.R., S.-J. Chiou, W.J. Silva, and J.R. Humphrey, 1997. Strong Ground Motion Attenuation Relationships for Subduction Zone Earthquakes, *Seism. Res. Lett.*, **68,** 58-73.

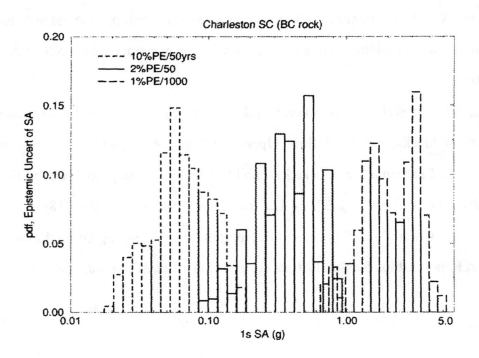

Figure 1. Conditional probability density functions for uncertainty in 1-s spectral acceleration for a given exceedance probability, computed at a site in Charleston, SC. *Left*, for PE 10% in 50 years, *center*, for PE 2% in 50 years, and *right*, for PE 1% in 1000 years.

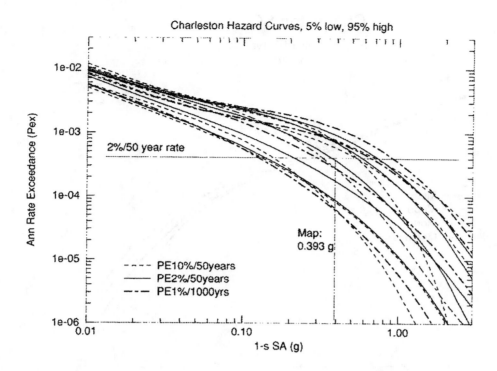

Figure (2a). 5%, 15%, 50%, 85%, and 95% fractile hazard curves determined at specific probabilities of exceedance for 1-s SA at Charleston, SC. *Dashed* curves, for PE 10% in 50 years, *solid* curves, for PE 2% in 50 years, and *dot-dash* curves, for PE 1% in 1000 years. Conventionally, *the* median hazard curve is spliced from the median hazard curves determined at each PE.

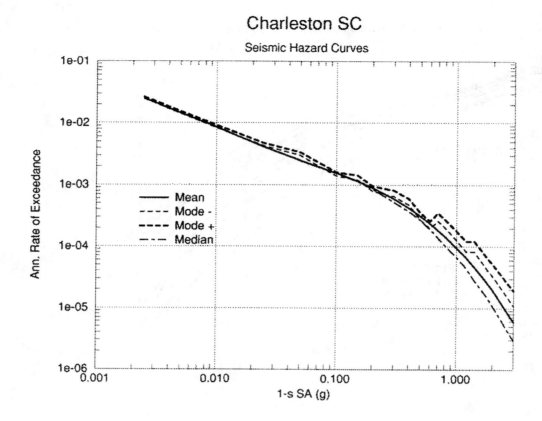

Figure (2b). Mean, modal, and median (50% fractile) hazard curves for 1-s SA at Charleston, SC. The mode is shown as a lower and upper value. Here, the conditional mean, median, or modal rate is computed at a fixed spectral acceleration.

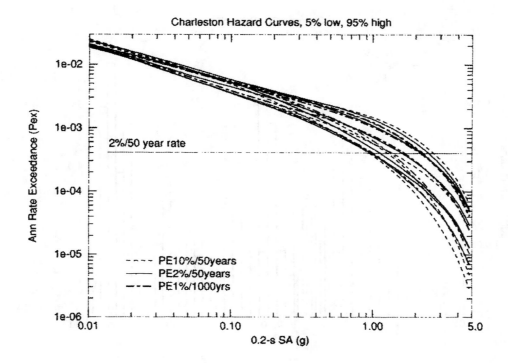

Figure 3. 5%, 15%, 50%, 85%, and 95% fractile hazard curves determined at specific probabilities of exceedance for 0.2-s or 5-Hz SA at Charleston, SC. *Solid* curves, for PE 10% in 50 years, *dashed* curves, for PE 2% in 50 years, and *dotted* curves, for PE 1% in 1000 years.

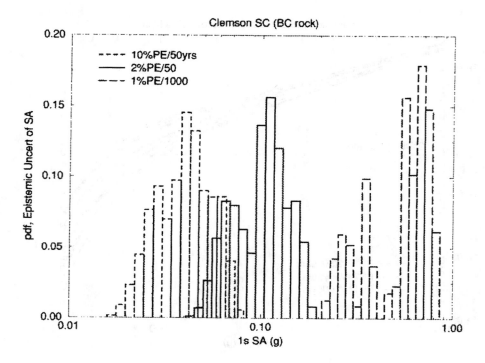

Figure 4. Probability density functions for uncertainty in 1-s spectral acceleration at a site in Clemson, SC. *Left*, for PE 10% in 50 years, *center*, for PE 2% in 50 years, and *right*, for PE 1% in 1000 years.

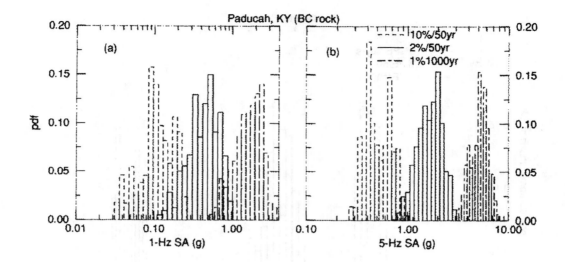

Figure 5. *(a)*, probability density functions for uncertainty in 1-s spectral acceleration at a site in Paducah, KY. *(b)*, pdfs for uncertainty in 0.2-s spectral acceleration at a site in Paducah, KY. *Left* distribution for PE 10% in 50 years, *center*, for PE 2% in 50 years, and *right*, for PE 1% in 1000 years.

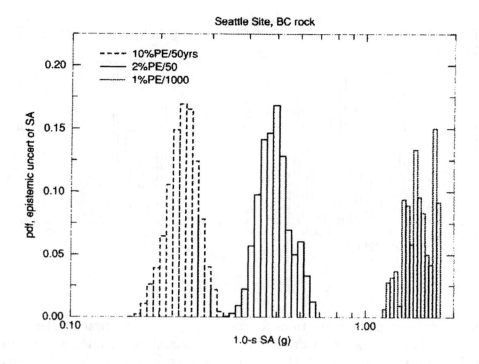

Figure 6. Probability density functions for uncertainty in 1-s spectral acceleration at a site in Seattle, WA. *Left*, for PE 10% in 50 years, *center*, for PE 2% in 50 years, and *right*, for PE 1% in 1000 years.

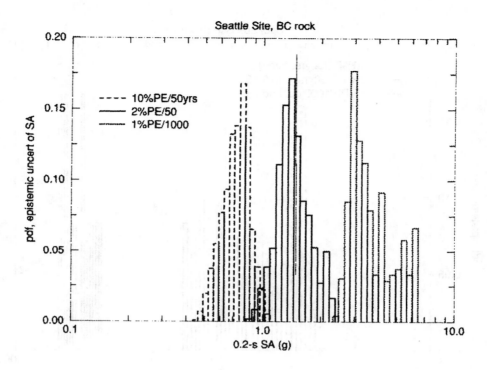

Figure 7. Probability density functions for uncertainty in 0.2-s spectral acceleration at a site in Seattle, WA. *Left*, for PE 10% in 50 years, *center*, for PE 2% in 50 years, and *right*, for PE 1% in 1000 years.

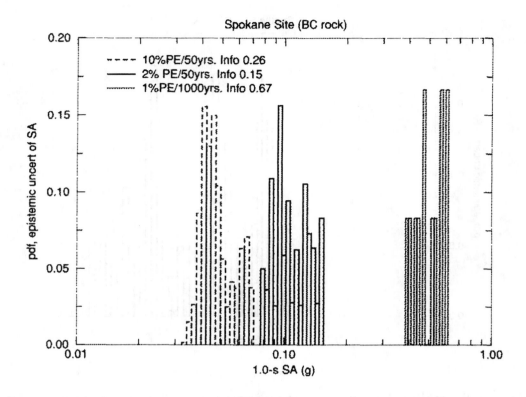

Figure 8. Probability density functions for uncertainty in 1-s spectral acceleration at a site in Spokane, WA. *Left*, for PE 10% in 50 years, *center*, for PE 2% in 50 years, and *right*, for PE 1% in 1000 years.

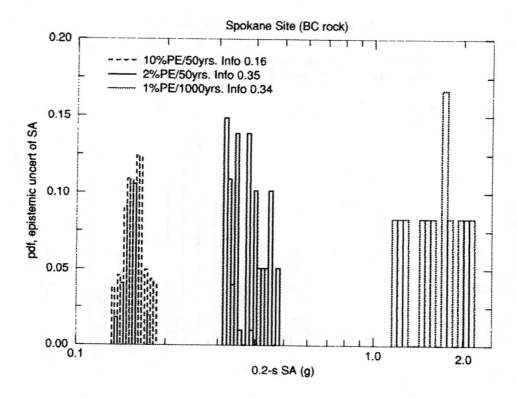

Figure 9. Probability density functions for uncertainty in 0.2-s spectral acceleration at a site in Spokane, WA. *Left*, for PE 10% in 50 years, *center*, for PE 2% in 50 years, and *right*, for PE 1% in 1000 years.

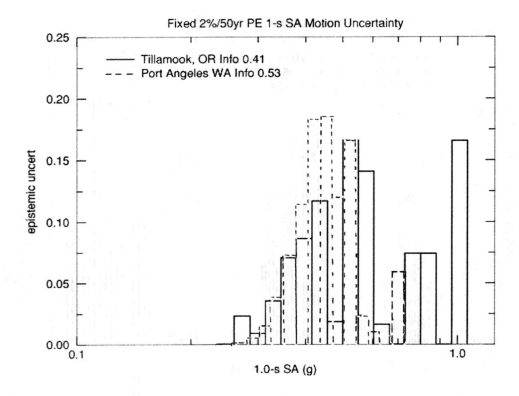

Figure 10 (a) Probability density functions for uncertainty in 1-s spectral acceleration at a site in Tillamook, OR (*solid*) and Port Angeles, WA (*dashed*) for PE 2% in 50 years.

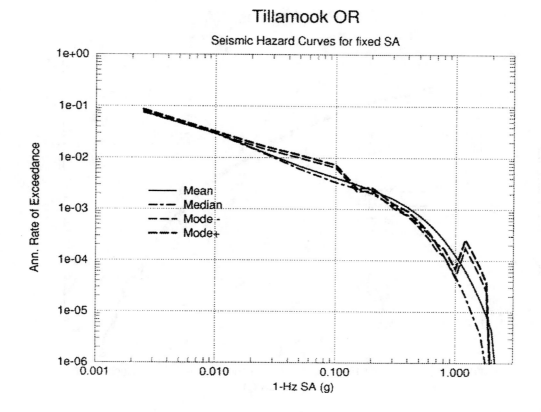

Figure 10(b) Seismic hazard curves for Tillamook for the estimated mean, median, and mode. The mode is shown as a lower and upper value. Here, the conditional mean, median, or modal rate is computed at a fixed spectral acceleration. (c) Seismic hazard curves for Port Angeles for the estimated mean, median, and mode, again computed at a fixed spectral acceleration.

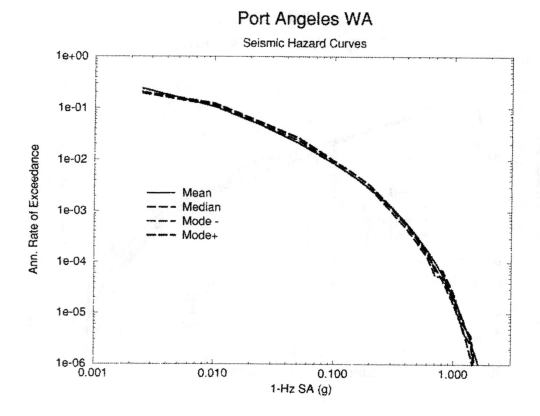

Figure 10(c) Seismic hazard curves for Port Angeles for the estimated mean, median, and mode, again computed at a fixed spectral acceleration.

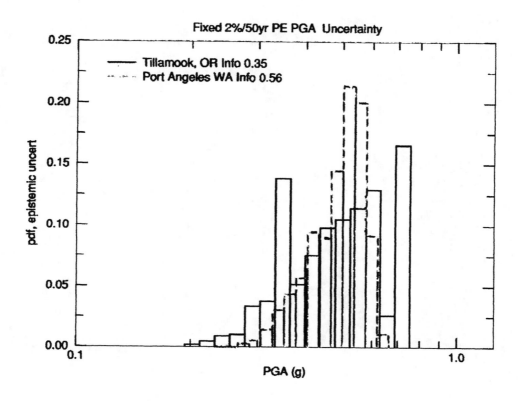

Figure 11. Probability density functions for uncertainty in peak ground acceleration at a site in Tillamook, OR (*solid*) and Port Angeles, WA (*dashed*) for PE 2% in 50 years.

SYSTEMS EVALUATION ISSUES

The Regional Economic Cost of a Tsunami Wave
Generated by a Submarine Landslide Off of Palos Verdes, California

Jose Borrero[1], Sungbin Cho[2], James E. Moore, II A.M. ASCE[3], and
Costas Synolakis[4]

Abstract

Recent developments in modeling of tsunami waves and economic impact analysis
are combined with data from recent offshore mapping of the Santa Barbara channel
and other locations to model the mechanism and economic impact of a tsunamigenic
undersea landslide in the vicinity of Los Angeles.

Introduction

The seismic sensitivity of the Los Angeles metropolitan region is well recognized,
Fortunately, severely damaging earthquakes have been relatively infrequent during
most of the past 200 years in the densely populated regions of coastal Southern
California, including the Los Angeles Basin, and the Santa Barbara -- Ventura
regions. See Figure 1. Several recent moderate earthquakes however, such as the
1994 M_w 6.7 Northridge earthquake and the 1987 M_w 6.0 Whittier Narrows
earthquake, have brought to light the hazard associated with thrust and reverse
faulting beneath Southern California (Dolan et al. 1995). There have been several
smaller, less damaging thrust and reverse earthquakes in the near shore region that
illustrate the possibility of a larger earthquake offshore. The shaking from an
earthquake of magnitude 7 or greater on an offshore thrust or reverse fault would
undoubtedly be damaging to coastal communities, and its effect would be enhanced
by its potential for generating a damaging tsunami.

The hazard to metropolitan Southern California posed by locally generated
tsunamis has received considerably less study than the hazards posed by onshore
earthquakes. This is likely to change. The mechanisms that generate tsunamis have
received considerable study following the unusually large waves associated with the
July 17, 1998 Papua New Guinea (PNG) tsunami. As a result of this increasing
scientific scrutiny, Southern California's susceptibility to tsunami damage is only
recently becoming understood.

Several locally generated tsunamis have been recorded in the region during
the past 200 years. One of the first large earthquakes to be recorded in Southern

---

[1] Research Assistant Professor of Civil Engineering, University of Southern California, KAP 210 MC-
2531, Los Angeles, CA, 90089-2531; phone 213-740-5129; jborrero@usc.edu.
[2] PhD Candidate, Urban Planning, School of Policy, Planning, and Development, University of
Southern California; and ImageCat, Inc., 400 Oceangate, Suite 1050, Long Beach, CA 90802; phone
562-628-1675; sungbinc@rcf.usc.edu, sc@imagecatinc.com.
[3] Professor of Civil Engineering; Public Policy and Management; and Industrial and Systems
Engineering, University of Southern California, KAP 210 MC-2531, Los Angeles, CA, 90089-2531;
phone 213-740-0595; jmoore@usc.edu.
[4] Professor of Civil Engineering; University of Southern California, KAP 210 MC-2531, Los Angeles,
CA, 90089-2531; phone 213-740-0613; costas@usc.edu.

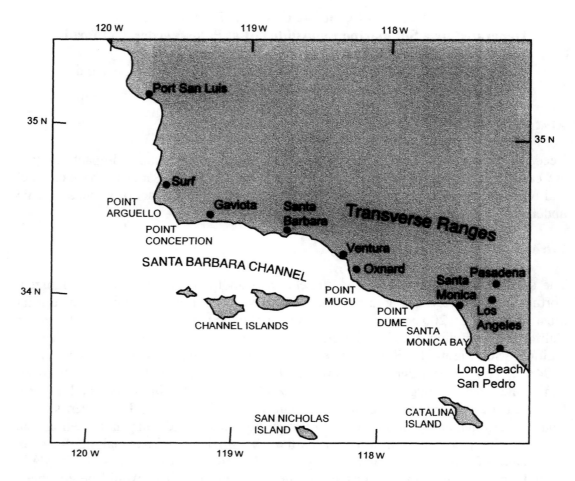

Figure 1: Southern California location map for tsunami and seismic activity over the past 200 years.

California, the December 21, 1812 Santa Barbara earthquake, appears to have generated a moderate tsunami that affected over 60 *km* of the Santa Barbara coast (Toppozada et al. 1981; Lander and Lockridge, 1989). Table 1 summarizes the details of several near-shore earthquakes, some of which generated tsunamis. There will be other such events, and this research focuses on the mechanisms and the potential economic impact of such a large, local tsunami event.

The Role of Research

Figures for 1994 show that the metropolitan Los Angeles area is responsible for nearly $746 billion of economic productivity annually (Cho, et al. 2001). Natural disasters such as earthquakes, fires, floods, and landslides act as agents of economic loss. Quantifying the economic impact associated with natural disasters has long been of research interest to economists, social scientists and engineers, but progress has been slowest in the social sciences. Much of the research on tsunamis has been in the engineering and geological fields. Progress in economic impact research is

Table 1: Possible Locally-Generated Tsunamis Along the Southern California Coast

Date (UTC)	Magnitude[a]	Area Affected	Waves Reported
1806 May 25	?	Santa Barbara	Boats beached(?)
1812 Dec. 21	7-7½ (M_I)	Santa Barbara Channel	3- to 4-m run-up at Gaviota
1854 May 31	?	Santa Barbara	Local inundation near embarcadero
1854 July 24	?	San Diego	0.1 m amplitude, 36 minutes period
1855 July 11	6 (M_I)	San Juan Capistrano	Two unusually heavy sea waves
1862 May 27	6.2 (M_I)	San Diego Bay	< 1-m
1878 Nov 22	?	Central Coast	Damage at Point Sal, Cayucos, Avila
1879 Aug 10	?	Santa Monica Bay	?
1895 Mar 9	?	San Miguel Island	Disturbed water; coastal landslide(?)
1927 Nov 4	7.3 (M_S)	Point Arguello	1.8-m run-up at Surf, 1.5-m run-up at Port San Luis
1930 Aug 31	5.25 (M_W)	Santa Monica Bay	Local oscillations to about 0.6 m
1933 Mar 11	6.25 (M_W)	Long Beach	Uncertain records
1979 Jan 1	5.0 (M_L)	Santa Monica Bay	Local oscillations (?); no tsunami
1989 Jan 19	5.0 (M_L)	Santa Monica Bay	Local oscillations (?); no tsunami

Sources: Modified and updated from McCulloch (1985), Lander et al. (1993), McCarthy et al. (1993), and Toppozada et al. (2000).

Note: a. Magnitude Scale: M_W = Moment Magnitude; M_S = Surface Wave Magnitude; M_L = Local (Richter) Magnitude; M_I = Seismic Intensity Magnitude; ? = Magnitude Not Given or Unknown.

recent. The physical science of earthquakes and tsunamis is challenging, but it is at least as diffitcult to explore the social impacts of disasters. Consequently, it is not surprising that there has been only limited previous attention to the socioeconomic impacts of natural disasters. This reflects a lag in social science research in this area.

Tsunamis, while generally related to earthquakes, have never been modeled quantitatively in terms of their potential economic impact. McCulloch (1985) estimated that tsunamis have been responsible for 0.2% of the cost of total earthquake damage between 1812 and 1964. This number, however, is derived primarily from the $32.2 million dollars (adjusted to 1983) of tsunami damage incurred in Crescent City, California after the 1964 Gulf of Alaska Earthquake. California has not suffered any tsunami damage since the 1964 event. However, coastal development has increased dramatically since then, placing at risk billions of dollars of additional property, businesses and infrastructure.

The core purpose of applied research in the natural disaster field is to assist policy makers. What types of information are required for creation of cost-effective policies? The large expenditures that are involved in many proposed mitigation programs suggest that a careful analysis of trade-offs is required. This means that analysts should study the full costs and benefits of prospective mitigation measures.. The benefits of mitigation are the costs avoided by the particular measure. Yet a discussion of costs avoided depends on analysts' ability to determine full costs. Standard engineering estimates of costs are most often restricted to replacement or repair costs. This focus on stocks is deficient from a policy perspective, because replacement cost estimates do not account for the opportunity costs resulting from damage to facilities. These latter costs are flows. These cascading economic impacts are incident to many economic sectors, including sectors in which facilities have not been damaged. Social science research can, therefore, make a substantial contribution by identifying a more complete representation of expected full costs with and without various proposed mitigations.

Some of the previous social science-based research on natural disasters has focused on measuring the total economic impacts of structure and contents damage. Work by Gordon, Richardson, and Davis (1998) on the business interruption effects of the 1994 Northridge earthquake indicates that an exclusive focus on structural damage ignores 25 - 30 percent of the full costs. Their analysis also identifies the geographical distribution of these impacts on individual cities and other small area zones. In 1994, business interruption job losses were estimated to be 69,000 person-years of employment, about half of which were jobs outside the area that experienced structural damage. Table 2 reports direct business interruption costs, i.e., opportunity costs resulting from lost production due to damage and reduced access to labor; and indirect and induced business interruption costs, which are incident to suppliers and households not experiencing damage. Reports that focus on replacement costs and fail to account for business interruption costs of such magnitudes substantially underestimate of the full costs of the event. Table 2 reports these business interruption costs by aggregate location, but much greater spatial detail is possible. This is useful, because the spatial details of such costs can be translated into household income strata (Gordon et al. 1998; Gordon, Moore, and Richardson 2002), providing insight into which income groups are most likely to benefit from various mitigation measures.

The Southern California Planning Model (SCPM)

The most widely used models of regional economic impacts are versions of inter-industry models. These attempt to trace all intra- and interregional shipments, usually at a high level of industrial disaggregation. Being demand driven, they only account for losses via backward linkages.

The Southern California Planning Model version 1 (SCPM1) was developed for the five-county Los Angeles metropolitan region, and has the unique capability to allocate all impacts, in terms of jobs or the dollar value of output, to 308 sub-regional zones, mostly municipalities. This is the result of an integrated modeling approach that incorporates two fundamental components: input-output and spatial allocation. The approach allows the representation of estimated spatial and sectoral impacts corresponding to any vector of changes in final demand. Exogenous shocks treated as changes in final demand are fed through an input-output model to generate sectoral impacts that are then introduced into the spatial allocation model.

The first model component is built upon the Regional Science Research Corporation input-output model. This model has several advantages. These include

- a high degree of sectoral disaggregation (515 sectors);
- anticipated adjustments in production technology;
- an embedded occupation-industry matrix enabling employment impacts to be identified across ninety-three occupational groups (This is particularly useful

Table 2: Business Interruption Losses from the 1994 Northridge Earthquake
(Jobs in person-years, Output in thousands of 1994 $)

Area	Direct		Indirect and Induced		Total	
	Jobs	Output	Jobs	Output	Jobs	Output
Impact Zone Total	34605.4	3,117,528	1,904.9	209,591.1	36,510.1	3,327.119.4
Rest of LA City	0.0	0.0	2,119.9	232,021.2	2,119.9	232,021.2
Rest of LA County	0.0	0.0	10,668.2	1,067914.1	10,668.2	1,067914.1
Rest of Region	0.0	0.0	8,260.7	877,532.0	8,260.7	877,532.0
Regional Total	34,605.4	3,117,528	22,953.7	2,387,058.5	57,559.1	5,504,586.9
Rest of the World	11,454.4	1,031,901.9	Not Computable		11,454.4	1,031,901.9
Total	46,059.8	4,149,430.3	22,953.7	2,387,058.5	69,013.5	6,536,488.8

Source: Gordon, Richardson, and Davis (1998).

for disaggregating consumption effects by income class and facilitates the estimation of job impacts by race.);

- an efficient mechanism for differentiating local from out-of-region input-output transactions via the use of Regional Purchase Coefficients (RPC); and
- the identification of state and local tax impacts.

The second basic model component is used for allocating sectoral impacts across 308 geographic zones in Southern California. The key was to adapt a Garin-Lowry style model for spatially allocating the induced impacts generated by the input-output model. The building blocks of the SCPM1 are the metropolitan input-output model, a journey-to-work matrix, and a journey-to-nonwork-destinations matrix. This is a journey-from-services-to-home matrix that is more restrictively described as a "journey-to-shop" matrix in the Garin-Lowry model.

The journey-from-services-to-home matrix includes any trip associated with a home based transaction other than the sale of labor to an employer. This includes retail trips and other transaction trips, but excludes nontransaction trips such as trips to visit friends and relatives. Data for the journey-from-services-to-home matrix includes all of the trips classified by the Southern California Association of Governments as home-to-shop trips, and a subset of the trips classified as home-to-other and other-to-other trips.

The key innovation associated with the SCPM1 is to incorporate the full range of multipliers obtained via input-output techniques to obtain detailed economic impacts by sector and by submetropolitan zone. The SCPM1 follows the principles of the Garin-Lowry model by allocating sectoral output (or employment) to zones via a loop that relies on the trip matrices. Induced consumption expenditures are traced back from the workplace to the residential site via a journey-to-work matrix and from the residential site to the place of purchase and/or consumption via a journey-to-services matrix. See Richardson et al. (1993) for a further summary of SCPM1.

Incorporating the Garin-Lowry approach to spatial allocation makes the transportation flows in SCPM1 exogenous. These flows are also relatively aggregate, defined at the level of political jurisdictions. With no explicit representation of the transportation network, SCPM1 has no means to account for the economic impact of changes in transportation supply. Tsunamis are likely to induce such changes, including capacity losses that will contribute to reductions in network level service and increases in travel delays. SCPM1 does not account for such changes in transportation costs, underestimating the costs of a tsunami.

We focus on a credible, hypothetical tsunami. Modeling the degree of inundation defines the lengths of time for which firms throughout the region will be non-operational. This allows the calculation of exogenously prompted reductions in demand by these businesses. These are introduced into the inter-industry model as declines in final demand. The I/O model translates this production shock into direct, indirect, and induced costs, and the indirect and induced costs are spatially allocated in terms consistent with the endogenous transportation behaviors of firms and household.

Implementing this approach is a data intensive effort that builds on the data resources assembled for SCPM1. In this case, results of structure damage to businesses are used to drive SCPM2. SCPM2 is a more advanced version of the Southern California Planning Model that endogenizes traffic flows by including an explicit representation of the transportation network. SCPM2 results are computed at the level of the Southern California Association of Governments' (SCAG) 1,527 traffic analysis zones, and then aggregated to the level of the 308 political jurisdictions defined for SCPM1. These jurisdictional boundaries routinely cross traffic analysis zones. Results for traffic analysis zones crossed by jurisdictional boundaries are allocated in proportion to area. Like SCPM1, SCPM2 aggregates to 17 the 515 sectors represented in the Regional Science Research Corporation's PC I-O model Version 7 (Stevens, 1996) based on the work of Stevens, Treyz, and Lahr (1983). Treating the transportation network explicitly endogenizes otherwise exogenous Garin-Lowry style matrices describing the travel behavior of households, achieving consistency across network costs and origin-destination requirements. SCPM2 makes distance decay and congestion functions explicit. This allows us to endogenize the spatial allocation of indirect and induced economic losses by endogenizing choices of route and destination. This better allocates indirect and induced economic losses over zones in response to direct tsunami losses to industrial and transportation capacity. See Cho et al. (2001) for a further summary of SCPM2.

The Geological Framework

The offshore region from Point Conception south to central Baja California is known as the Southern California Borderlands. See Figure 2. The Borderlands are a geologically complex region comprised of different tectonic regimes, complicated bathymetry consisting of deep basins, towering ranges and steep walled canyons. The

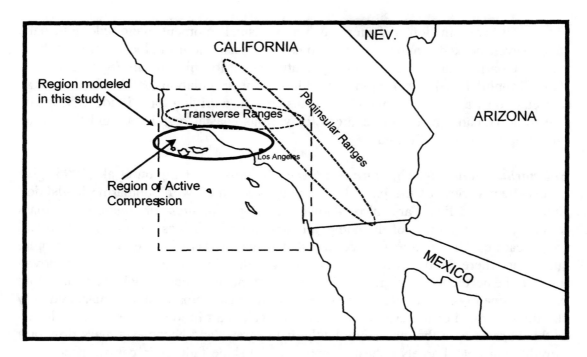

Figure 2: Compressional tectonics offshore of Southern California.

steep topography that is visible on land does not stop at the water's edge, but rather continues under the sea.

Southern California lies astride a major transition between two tectonic provinces. The region to the south is dominated by northwest-trending, strike-slip faults, whereas the area to the north is characterized by west-trending mountain ranges - the Transverse Ranges - that have developed above west-trending reverse, oblique reverse, and left-lateral strike-slip faults. Where these thrust systems extend offshore, they may represent significant potential sources of tsunamis.

Tsunamis generated by Submarine Landslides. The standard paradigm of large tsunamis generated primarily by tectonic uplift or subsidence has come under increasing scrutiny as a result of the unusually large waves associated with the July 17, 1998 Papua New Guinea (PNG) tsunami. The Papua New Guinea tsunami was generated after a relatively small earthquake with an approximate $M_w = 7.0$ (Kawata et al. 1999, Matsuyama et al. 1999). The large runup values and extreme inundation, as well as the devastation of coastal communities observed along the PNG coast prompted even the scientists in the International Tsunami Survey Team (ITST) to rethink standard models and search for an alternate explanation for the cause of the wave (Kawata et al. 1999). The runup distribution plotted along the PNG coast showed a very large peak that tapered off rapidly within 10 *km* of the peak. This distribution is unlike that observed in the near field from classic subduction zone earthquakes generating tsunamis, such as the 1995 Jalisco-Colima earthquake (Borrero et al. 1995).

The combination of factors such as the small moment magnitude, unusually large waves, peaked runup distribution plus other seismological clues suggested that a giant submarine mass failure was the causative agent for this tsunami (Synolakis et al. 2002, Tappin et al. 1999, Tappin et al. 2001). The speculation that the PNG tsunami was caused by a submarine mass failure has become a driving force in identifying non - tectonic tsunami generation sources for various coastal regions around the world, including Southern California.

Submarine Landslides Offshore Southern California. Hampton et al. (1996) give an excellent review of the basic terminology associated with submarine landslides. Submarine landslides are also known as slope failures or underwater mass movements. Regardless of the name, all submarine landslides possess the same two basic features, the rupture surface and a displaced mass of material. The rupture surface is where the downslope motion originated. The displaced material is moved through the acceleration of gravity along a failure plane. There may be several failure planes in one mass movement. The displaced mass can remain largely intact and only slightly deformed or the displaced mass can break apart in to separate sliding blocks. In the extreme case the mass completely disintegrates and becomes a mass flow or a turbidity current. Largely cohesive or block like slope failures in canyon heads may evolve into or trigger turbidity currents in basins (Gorsline 1996).

Submarine landslides are of two general types, the rotational slump and the translational slide. When the rupture surface cuts through a homogeneous material, is scoop-shaped and concave upward, the sliding mass follows a circular arc. This type of slide is known as a *rotational slump*. If the rupture surface is more or less planar and the failure plane is the result of material inhomogeneities, i.e., bedding planes, the motion of the displaced mass is translational and is called a *translational slide*. A series of consecutive failures that propagate upslope is called a *retrogressive* failure (Hampton et al. 1996).

Such submarine landslides can occur in a wide range of sizes, over several orders of magnitude ranging from very small to enormous. As a matter of comparison, Hampton et al. (1996) note that the largest documented subaerial slide might contain tens of cubic kilometers of displaced material whereas one submarine slide mapped off South Africa might contain a displaced volume of over 20,000 km^3.

The near offshore region from Point Conception to the Mexican Border is characterized by a mainland shelf that runs from the shoreline to depths of 70 to 100 m. This shelf varies in width from 3 to 20 km. It is narrowest in the southern reaches of this zone and broadest in Santa Monica Bay and immediately south of the Palos Verdes Peninsula. The shelf has a relatively gentle slope on the order of a few degrees that runs to a slope break, and then a steeper 5° to 15°slope that drops off into deeper offshore basins of 800 m depths or more. The shelf is periodically cross cut by deep canyons along its entire extent. Starting in the north, the major canyons are Hueneme, Mugu, Dume, Santa Monica, and Redondo. South of the Palos Verdes peninsula there are 5 more major canyons, San Gabriel, Newport, Carlsbad, La Jolla and Coronado (Clarke et al. 1985, McCarthy, 1993, Synolakis et al. 1997b).

The slopes offshore of Southern California generally have thick accumulations of under-consolidated sediment of Quaternary age (Clarke et al. 1985).

These water-saturated sediments generally have a lower shear strength than comparable onshore sediments. Therefore underwater slope failures are generally larger and occur on lower slopes than on land (Clarke et al. 1985). Decadal and generational floods discharge sediment in amounts one to three orders of magnitude larger than the average annual contribution. These events load the shelf and canyons with material. Seismic activity can trigger slope failures where sedimentation is high and sediments are unstable (Gorsline, 1996). Clarke et al. (1985) note that most offshore slope failures appear to be composite rather than single events. They also mention that it is difficult, if not impossible, to determine the timing and rate of motion of individual slides. They go on to say that " zones of past failure should be viewed as having an unknown potential for renewed movement," in that "some may be more stable than unfailed accumulations of sediment on the adjacent slopes; others, however, may have unchanged or even reduced stability."

Clarke et al. (1985) discuss specific areas of submarine landslides offshore Southern California. Their comprehensive list mentions every offshore canyon, slope and headland from the Santa Barbara Channel to San Diego County. Modeling potential tsunamis from these types of slides is not straight forward, especially since there is no indication whether or not these features were generated as single catastrophic events or through slow gradual movements over time. Nonetheless their morphology is intriguing and further research is necessary to determine the age of these features and the details of their motions.

Scenario Event for this Study. For this study, a tsunami scenario based on waves being generated by a submarine landslide offshore of the Palos Verdes Peninsula is used. A landslide source is chosen for two reasons. First, in the period between 1992 and 2001 several locally generated and destructive tsunamis have been associated with submarine or subaerial landslides. The 1998 Papua New Guinea tsunami was responsible for over 2000 deaths and is believed to have been caused by a large (4 km^3) offshore slump (Tappin et al. 1999, Kawata et al. 1999, Tappin et al. 2001, Synolakis et al. 2002). In August of 1999, a large earthquake near Istanbul Turkey caused significant landslides and slumping along the shores of the Sea of Marmara and contributed to waves, which damaged port facilities. In September of 1999, a large rockfall on the south shore of Fatu Hiva in the Marquesas Islands caused tsunami runup in excess of 2 *m* which inundated a local school and nearly killed several children (Okal et al. 2002a). In December of the same year on Pentecost Island, Vanuatu a highly localized tsunami wave with runup of up to 6 *m* wiped out an entire village and killed 2 people (Caminade et al. 2000).

Second, recent offshore mapping work has found evidence of significant sliding and slumping offshore of Southern California, particularly in the Santa Barbara Channel (Greene et al. 2000) and off of the Palos Verdes Peninsula (Bohannon and Gardner, 2003). Studies have shown that these events could have been tsunamigenic with wave heights ranging from 5 to 20 *m* (Borrero et al. 2001, Bohannon and Gardner 2003, Locat et al. 2003). Furthermore, due to the proximity to a major port, a tsunami generated off of Palos Verdes would have the greatest impact on economic activity.

Quantifying the Innundation Effects of a Tsunami Disaster

The Palos Verdes Slide. Through high resolution bathymetric surveys, Bohannon and Gardner (2003) identified a potentially tsunamigenic submarine landslide feature on the San Pedro escarpment south west of the Palos Verdes Peninsula. This feature appears in Figure 3. The thin black contours are bathymetry and topography repectively. Locat et al. (2003) analyzed the mobility of this feature and concluded that the state of the mapped debris field implies that the Palos Verdes Slide must have moved as a large block in a single catastrophic event.

Bohannon and Gardner (2003) and Locat et al. (2003) all modeled the waves generated by this Palos Verdes Slide (PVS). Bohannon and Gardner used an energy scaling relationship for a rock avalanche with the following dimensions derived from their surveys: Thickness (T) = 70 m and Length (L) = 4000 *m*. They assumed a 2% transfer potential energy to tsunami wave energy as the sliding mass fell 350 m. Their analysis resulted in tsunami wave amplitude of 10 *m* for a landslide density of 1500 *kg/m³*.

Locat et al. (2003) also proposed a tsunami wave generation mechanism based on the energy equation of Murty (1979). They proposed initial wave heights of 10 to 50 *m* depending on the value of Murty's μ, "the energy transfer efficiency," which they varied from 0.1% to 1% (.001 to .01).

It is important to note that both of these amplitudes are estimated for a solitary wave generated by a transfer of energy from the sliding mass in to tsunami waves. For our purposes, we model the initial wave as an asymetrical dipole. Our model is based on curve fits to laboratory data developed by Watts (1998, 2000). The inital wave is shown as the thick, solid, concentric contours on Figure 3. It has a negative amplitude of 12 *m* and a positive amplitude of 4 *m*. This analysis is consistent with the methods used by Bohannon and Gardner (2003) and Locat et al. (2003).

Modeling Runup. For this study, the same conditions were used to generate a single case inundation map for a landslide-generated wave off of the Palos Verdes Peninsula. Two cases are modeled, one with the Port of Los Angeles breakwater in place and one without the breakwater. Comparing results shows the effect of the narrow openings between breakwater segments. Modeling suggests current velocities of over 3 *m/sec* in these openings. Instantaneous peak velocities are modeled to be over 10 *m/s*, but generally occur on the steep cliffs of Palos Verdes, west of the entrance to the Ports. The computed runup in the region of the ports is not affected by the presence of the breakwater, because the wave length of the tsunami is relatively long and breakwaters are generally invisible to long waves.

Figure 3 shows the initial wave and the local bathymetry and topography. Figures 4a,b and show the time series of water levels at locations indicated in Figure 3. The maximum drawdown reaches the outer harbor area about 6 minutes after wave generation, with the maximum positive wave arriving one minute later. This illustrates the extremely short time that is available for emergency planners to deal with when considering the effects of landslide-generated tsunamis off of Southern California. Figures 5a,b provide plots of snapshots of the depth averaged velocities at various times.

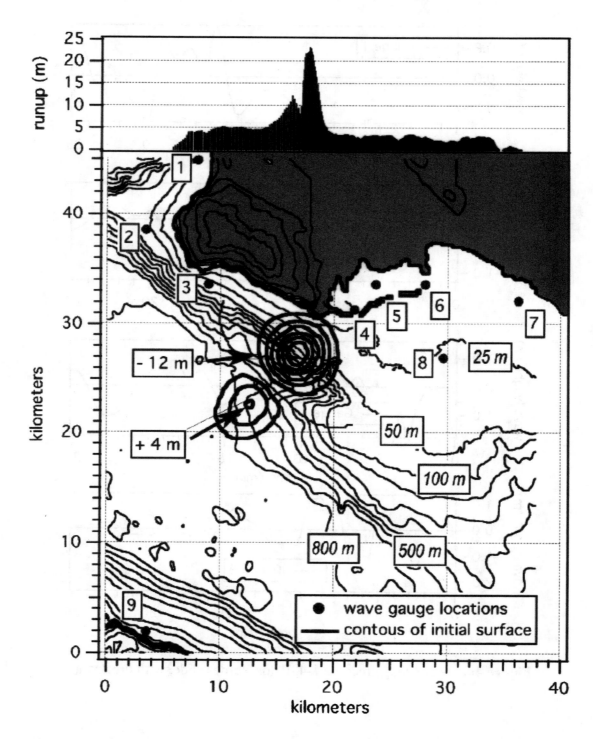

Figure 3: Initial wave used in the Palos Verdes slide simulations and runup along the south-facing shoreline boundary: Breakwater included. The asymetrical dipole has a negative amplitude of 12 m and a positive amplitude of 4 m. The thicker concentric contours are of the initial wave used in the simulations. The thin black contours are bathymetry and topography, respectively. Bathymetry contour values are in italics. Wave gauges are indicated by numbers.

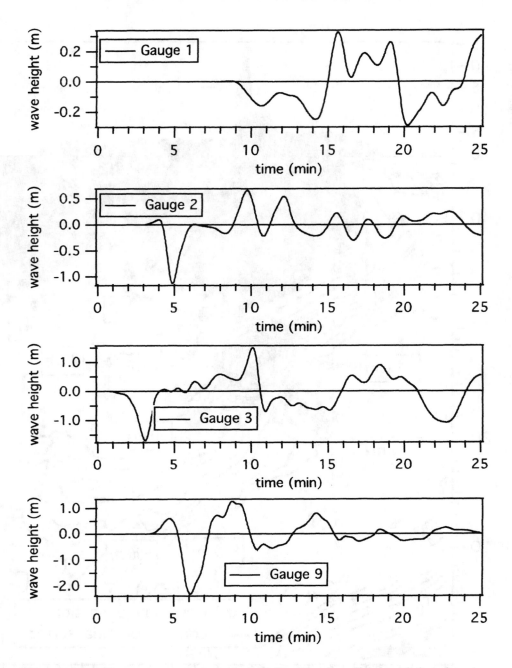

Figure 4a: Wave gauge records 1-4, Palos Verdes slide, with breakwater.

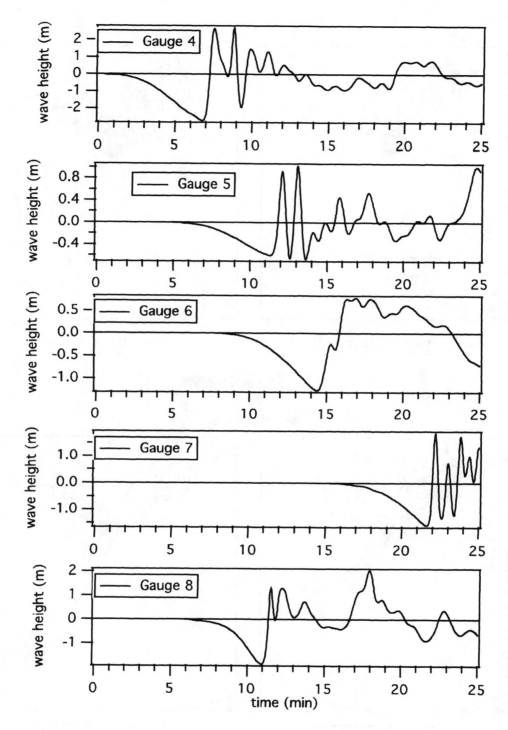

Figure 4b: Wave gauge records 5-8, Palos Verdes slide, with breakwater.

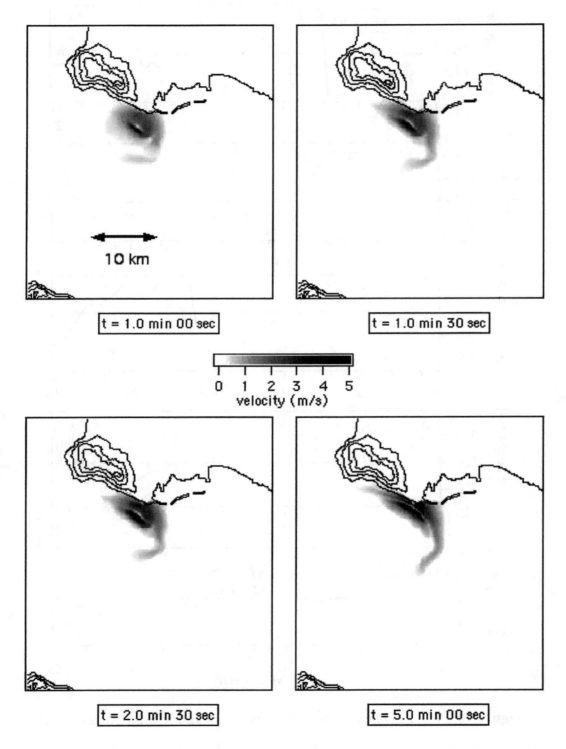

Figure 5a: Depth-averaged water velocities, t = 1.0 minute to t = 5 minutes, with breakwater.

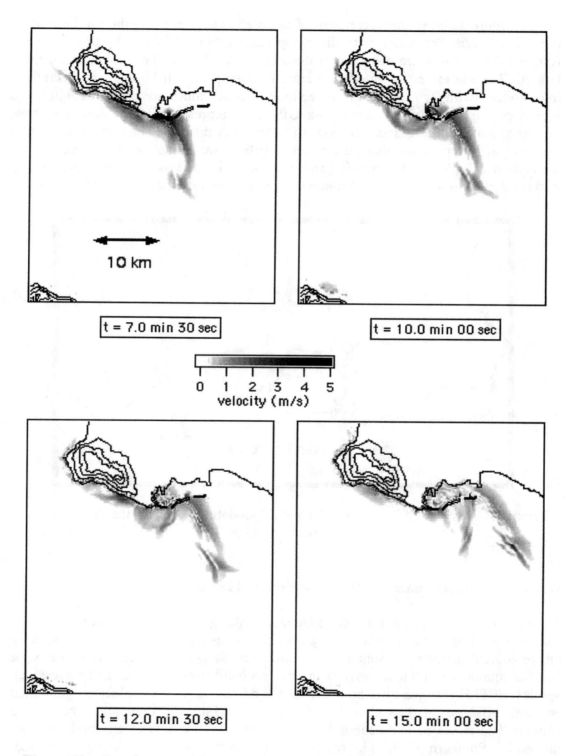

Figure 5b: Depth-averaged water velocities, t = 7.0 minutes 00 seconds to t = 15 minutes, with breakwater.

Figure 3 shows that large peak of up to 25 *m* is seen along the southern tip of the Palos Verdes Peninsula, with the runup value dropping of rapidly to either side. Runup values of 4 *m* are observed in the area of the Ports of Los Angeles and Long Beach. To determine the inundation area, the runup plot in Figure 3 is discretized into 4 zones with 2, 4, 6 and 20 *m* sections. These values are used in conjunction with topographic maps to generate a GIS layer representing the inundated area. Figure 6 shows these inundation zones plotted with different shadings representing the level of runup. Note that on the steep cliffs of the Palos Verdes Peninsula, the inundation is limited to the fringing shore, however in the low lying areas around the Ports and to the east, even 2 *m* of runup can produce significant inundation.

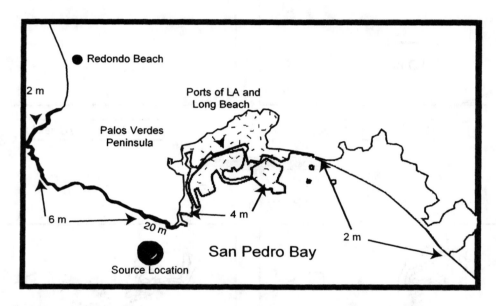

Figure 6: Discretized 2, 4, 6, and 20 meter inundation zones for the Palos Verdes Landslide tsunami.

Quantifying the Economic Effects of a Tsunami Disaster

The economic impact of a destructive tsunami striking a metropolitan area is the least studied aspect of tsunami hazard mitigation. Previous studies have focused primarily on geological aspects of tsunami generation mechanisms, expected tsunami wave heights, inundation and to a lesser extent, the probability of occurrence. This research applies SCPM2 to quantify potential costs related to a major locally generated tsunami offshore of Los Angeles, California. Tsunami inundation zones are converted to zones of lost economic productivity, and the effects are modeled over the entire Southern California region. The results of the modeling show that distributed losses related to a major tsunami in Southern California could exceed $7 billion, and be as high as $42 billion, depending largely on the degree to which the infrastructure at the Port of Los Angeles and the Port of Long Beach is affected.

Direct, Indirect, and Induced Losses. The inundation map shown in Figure 6 defines the inputs to the Southern California Planning Model 2 (Cho et al. 2001). As noted in Section 1, this spatial economic impact model discretizes the five-county Los Angeles Metropolitan region into 1,527 zones, and reports results for 308 zones corresponding mostly to municipalities and City of Los Angeles council districts. The model is able to calculate direct losses within a damaged area as well as the distributed economic effects of these losses throughout the regional economy. The model also calculates the distributed effect of damage to the transportation network. These quantities are fundamentally different from damage costs. Damage costs are the costs of repairing or replacing damaged or destroyed property. These replacement costs are not considered or quantified in this study. Such costs already receive considerable attention in the literature. Engineering estimates of direct costs are most often replacement costs. The term "direct cost" has a different meaning in the context of an input-output model. Direct losses arise from lost opportunities to produce, or, in the case of the port damage, to ship. Indirect and induced losses arise as people and businesses in the damaged areas become unable to work or generate income as a result of the event. Indirect losses occur to suppliers whose products and services are no longer purchased by damaged firms and households. Induced losses are losses incident to labor.

For this initial study, economic activity is assumed to stop for one year within the inundation zone. This serves for purposes of a first illustration, and is a simple assumption to adjust. Longer or shorter interruption periods can be scaled proportionate to these results, or a period of partial operation can be introduced to account for facilities' gradual return to service. Table 3 shows these values in each of the affected municipal zones represented in SCPM2. Note that the number of inundated zones is small relatively to the number of zone represented in the model. Direct losses accrue only in inundated locations, but indirect losses can accumulate through out the region. SCPM2 is used to establish a baseline representation of the regional economy, its outputs, and transportation costs. The direct losses associated with inundated locations are introduced into SCPM2 as a reduction in final demand for exports from affected economic sectors. SCPM2 calculates and allocates indirect and induced losses regionwide. Table 4 shows the indirect and induced losses incurred system-wide as a result of the damage cause by inundation. In total, these estimated losses account for 0.99% of the total economic output of the five-county Southern California region.

Impacts on production facilities are only part of the story. A tsunami in this location produces a special threat to the facilities at the Port of Los Angeles and the Port of Long Beach. These ports are of central importance to the regional economy, and the loss of transshipment capabilities at these sites would have a profound effect. In the worst case, export flows currently using seaport facilities would terminate so long as the ports were out of service. Table 5 gives the percentage of export flows using Port of Los Angeles / Port of Long Beach facilities within aggregate industrial sectors, and calculates the direct economic loss incurred by eliminating the production of these shares from the local economy. Table 6 shows the model estimates of indirect and induced losses resulting from the direct losses calculated in Table 5. These tables describe is a loose upper bound on the direct economic impact

Table 3: Direct loss and Annual Baseline Production in Inundated Areas

City	Baseline ($1000)	Direct Loss ($1000)	Direct Loss as a % of Baseline
Carson	6,591,962	85,736	1.30
Hawaiian Gardens	216,150	323	0.15
Long Beach	22,838,571	3,607,647	15.80
Palos Verdes Estates	416,315	32,338	7.74
Rancho Palos Verdes	510,586	26,903	5.27
Wilmington / San Pedro	5,675,587	314,931	5.55
Unincorporated LA County	17,623,822	2,565	0.01
Garden Grove	4,969,415	190	0.00
Huntington Beach	7,031,246	299,580	4.26
Los Alamitos	1,481,826	12,543	0.85
Rossmoor CDP	120,899	5,761	4.76
Seal Beach	1,398,293	103,892	7.43
Westminster	2,238,251	6,908	0.31
Unincorporated Orange County	3,401,272	3,051	0.09
Total	74,513,195	4,502,257	6.04

Table 4: Direct, Indirect and Induced Losses Throughout the Five-County Region

	Loss ($1000)	Loss as a % of Total Output[a]
Direct	4,502,257	0.60
Indirect	1,541,117	0.21
Induced	1,325,883	0.18
Total	7,369,257	0.99

Note: a. Total 1994 five-county output = $ 745,818.8 M.

Table 5: Maximum Direct Losses Due to Loss of Port Services.

Industry	Total Exports[a] ($ Millions)	Port Share of exports (%)	Direct Impact ($ Millions)
Mining	158.5	46.90	74.34
Durable	25,172.7	40.61	10,628.73
Non-Durable	37,595.9	23.23	8,732.27
Wholesale	19,394.3	13.05	2,531.60
Sum	82,321.4		21,966.94[b,c]

Notes: a. Total Exports from PC I-O Transaction Table.
   b. Total 1994 five-county output = $745,818.8 M.
   c. Ratio of Direct Impact to total output = 2.95%.

Table 6: Direct, Indirect and Induced Losses in port areas.

	Economic Impact ($ Millions)	Share of Baseline Total Output (%)[a]
Direct	21,966,941	2.95
Indirect	8,762,751	1.17
Induced	5,451,162	0.73
Total	36,180,854	4.85

of port damage. Direct impacts are defined by the period the port facilities are unavailable, and are an input to the model. This maximum input is one of the scenarios investigated below.

Transportation Impacts. Potential damage to the transportation infrastructure in Southern California implies additional impacts. In the case of a tsunami, inundation would affect surface streets and might not close elevated freeway segments. However, for the purposes of this study, freeway segments were assumed to be closed. Other assumptions, such as partial capacity on inundated facilities, are simple to accommodate. The SCPM2 representation of the transportation network includes freeways, state highways, and high design arterials. Small surface streets are not included in the model.

Some degree of export activity will be possible despite a port closure because the mode of transport can be switched from ship to truck or rail. However, this change may not occur in the short-term. These uncertainties suggest four analysis scenarios of increasing severity.

- Scenario 1:
  - Direct + indirect + induced business loss in the inundated area.
  - No freeway links are closed.
  - Ports Los Angeles and Long Beach are functional.
  - No reduction in export capabilities occurs.
- Scenario 2:
  - Direct + indirect + induced business loss in the inundated area.
  - Freeway links in the inundated area are closed for one year.
  - Ports Los Angeles and Long Beach are functional.
  - No reduction in export capabilities occurs.
- Scenario 3
  - Direct + indirect + induced business loss in the inundated area.
  - Freeway links in the inundated area are closed for one year.
  - Ports Los Angeles and Long Beach are closed for one year.
  - No reduction in export capabilities occurs because exported goods are transported by truck and rail rather than ship.
- Scenario 4:
  - Direct + indirect + induced business loss in the inundated area.
  - Freeway links in the inundated area are closed for one year.
  - Ports Los Angeles and Long Beach are closed for one year.
  - Export flows that used to be transported through the ports are now impossible. In addition to the 0.99% decrease in total economic activity implied by damage to production facilities in the inundation area, there is an addition 4.85% reduction in exports from all over the Southern California area.

Tables 7 to 10 summarize the results of these economic impact analyses. Table 7 shows the direct, indirect, and induced losses incurred as a result of tsunami inundation. Scenarios 1 through 3 all have the same impact in these separate loss categories, but different delay costs as a result of variable damage to the transportation network. Scenario 4 adds the direct, indirect, and induced costs associated with a closure of the ports. See Table 6.

Table 8 summarizes the costs of transportation delays incurred due to damage to the transportation network, and the diversion of port flows to the remainder of the road system. Delays are reported in terms of Passenger Car Unit (PCU) hours and $billions. Delay accumulates on an undamaged network, and to even greater degree on a damaged network. Table 9 gives the difference relative to baseline delay values for each scenario. Table 10 summarizes the total of the delay costs associated with interruption of the transportation services combined with the direct, indirect and induced costs associated with tsunami inundation of production and port facilities.

Table 7: Summary of Direct, Indirect and Induced Losses for Each Scenario

	Type of Loss			
	Direct Loss ($M)	Indirect Loss ($M)	Induced Loss ($M)	Total ($M)
Scenario 1	4,502.257	1,541.117	1,325.883	7,369.257
Scenario 2	4,502.257	1,541.117	1,325.883	7,369.257
Scenario 3	4,502.257	1,541.117	1,325.883	7,369.257
Scenario 4	26,469.198	8,903.868	677.045	43,550.111

Table 8: Summary of Transportation Network Delay Costs for Each Scenario

	Driver Delay		Freight Delay		Total Delay[b]	
	PCU[a] Hours	$ Billion	PCU Hours	$ Billion	PCU Hours	$ Billion[c]
Baseline	6,319,364	21.290	762,110	4.550	7,081,474	25.839
Scenario 1	6,323,171	21.302	756,912	4.518	7,080,083	25.821
Scenario 2	6,351,051	21.396	804,195	4.801	7,155,246	26.197
Scenario 3	6,380,809	21.497	852,092	5.087	7,232,901	26.583
Scenario 4	6,329,239	21.323	676,524	4.039	7,005,762	25.361

Notes: a. Passenger Car Units, including both cars and trucks.
b. 365 travel days per year, 1.42 passengers per car, and 2.14 Passengers Car Units per truck.
c. $6.5 per hour for individuals and $35 dollars per hour for freight.

Analysis

From the results above, it is clear that the costs associated with a tsunami disaster from a local tsunamigenic landslide source would include substantial direct, indirect, and induced costs associated with lost economic opportunity. This figure is on the order of $7 billion per year, and is separate from the replacement and repair cost of damaged facilities. Furthermore, damage to port facilities could produce much larger losses. If the loss of port services equates to the loss of export services, then the economic impact of the scenario tsunami is approximately $36 billion in losses. The greatest increase in transportation delays occurs in the case where port export flows

Table 9: Network Losses, Difference (Δ) from Baseline Flows

	Driver Delay		Freight Delay		Total Delay	
	PCU Hours	$ Billion	PCU Hours	$ Billion	PCU Hours	$ Billion
Scenario 1	3,806	12.824	-5,198[a]	-31.029[a]	-1,391[a]	-18.206[a]
Scenario 2	31,687	106.751	42,085	251.233	73,772	357.984
Scenario 3	61,445	207.006	89,982	537.158	151,427	744.163
Scenario 4	9,874	33.266	-85,586[a]	-510.917[a]	-75,712[a]	-477.651[a]

Note: a. Negative values represent reductions in network delay caused by reduced production in inundated areas or are due to loss of port access.

Table 10: Loss Totals

	Economic Loss	Network Loss	Total
Scenario 1	7,369.257	-18.206	7,351.051
Scenario 2	7,369.257	357.984	7,727.241
Scenario 3	7,369.257	744.163	8,113.420
Scenario 4	43,550.111	-477.651	43,072.460

are forced to switch from the waterways to land based routes, thus creating further congestion and delays on Southern California's transportation network.

Closer inspection of Tables 9 and 10 show some interesting effects. Negative values represent reductions in delay relative to baseline conditions. These reductions in delay are caused by reduced production in inundated areas or are due to loss of port access. However, this improved level of service for the travelers that remain constitutes a rather small positive impact in the face of overwhelming costs.

The difference between scenarios 3 and 4 is that in scenario 3 there is no accounting for the direct cost of damage to the closed ports. The increased losses in scenario 3 consist of increased transportation costs associated with shifting exports from sea to land-based modes. Scenario 4 is the extreme case: the ports are shut down and no exports are shifted to alternative modes. Production associated with these exports simply ceases, even in local facilities outside the inundation area. These losses range from $7 billion to $43 billion and provide upper and lower bounds for the economic impacts associated with this particular tsunami. Neither extreme is likely: Physical damage to wharves, piers and loading facilities would be expected to force some export flows to shift to other modes of transportation, but not all.

Finally the potential economic losses associated with damage to the ports outweigh the totals from the remainder of the inundated region by a factor of five. This figure alone demonstrates the vulnerability of the port infrastructure and pressing need for a comprehensive tsunami hazard assessment in major US shipping ports. This is particularly true along the tsunami prone US west coast, and extremely important in locations exposed to tsunamigenic landslide sources. This includes Southern California.

As a comparison, Cho et al. (2001) used the same method and baseline economic data to calculate economic loss associated with a hypothetical magnitude 7.1 earthquake on the Elysian Park blind thrust fault under downtown Los Angeles. They calculated that an earthquake of this type could produce as much as $135 billion in total costs, with a median amount of $102 billion. The scenarios defined here vary between 5 percent and 30 percent of their maximum estimate. However, it is important to remember that these tsunami costs could be incurred *in addition* to the earthquake costs described by Cho et al. (2001). Tsunamigenic landslides are a recently identified risk, but the tsunami risk from seismic sources remains undiminished.

Summary

The example presented here illustrates a methodology for calculating the economic impact of tsunami inundation. Instead of the standard step of merely quantifying the repair and replacement costs associated with tsunami damage, this method distributes the total economic effects of this damage to households and businesses throughout the metropolitan economy. Not all post-event economic behavior is knowable, but this approach makes it possible to calculate the economic impacts associated with a variety scenarios, including changes in export transshipment modes. The results of this preliminary study suggest that a devastating local tsunami could cause between $7 and $40 billion worth of direct, indirect and induced costs, and transportation related delays.

Acknowledgements

We thank Profs. Peter Gordon and Harry Richardson of the School of Policy, Planning and Development and the Department of Economics at the University of Southern California for their contributions to this research. We are also grateful for the contributions of two anonymous referees. Any errors or omissions remain the responsibility of the faculty authors.

This research was supported in part by the University of Southern California Zumberge Research Innovation Fund, in part by the Earthquake Engineering Research Centers Program of the NSF under Award Number EEC-9701568, and in part by the Institute for Civil Infrastructure Systems (ICIS) at New York University (in partnership with Cornell University, Polytechnic University of New York, and the University of Southern California) under NSF Cooperative Agreement No. CMS-9728805. Any opinions, findings, and conclusions or recommendations expressed in this document are those of the authors and do not necessarily reflect the views of the

National Science Foundation or of the Trustees of the University of Southern California.

References

[Bohannon and Gardner, 2003] Bohannon, R.G., and Gardner, J.V. (2003). Submarine landslides of San Pedro Sea Valley, southwest Los Angeles basin. *Marine Geology*, in press. Available from the authors.

[Boore and Stierman, 1976] Boore, D., and Stierman, D. (1976). Source Parameters of the Point Mugu, California Earthquake of February 21, 1973. *Bulletin of the Seismological Society of America*, V.66, (2), p. 385 - 404.

[Boarnet 1998] Boarnet, M., (1998). Business Losses, Transportation Damage and the Northridge Earthquake, *Journal of Transportation and Statistics*, V.1, p. 49-64.

[Borrero et al. 1995] Borrero.J., M. Ortiz, V. Titov, and C. Synolakis (1995). Field survey of Mexican tsunami produces new data, unusual photos. *Eos, Transactions of the American Geophysical Union* V. 78, p. 85, 87 - 88.

[Borrero et al. 2001] Borrero, J.C., Dolan, J.F., and Synolakis, C.E. (2001). Tsunamis within the eastern Santa Barbara Channel. *Geophysical Research Letters*, V.28 (4), p. 643-646.

[Byerly 1930] Byerly, P. (1930) The California Earthquake of November 4, 1927 *Bulletin of the Seismological Society of America*, V. 20, p. 53-60.

[Caminade et al. 2000] Caminade, P., Charlie, D., Kanoglu, U., Matsutomi, H., Moore, A., Rusher, C., Synolakis, C.E., and Takahashi, T. (2000). Vanuatu Earthquake and Tsunami Cause Much Damage, Fee Casualties. *Eos, Transactions of the American Geophysical Union*, V. 81 (52), p. 641, 646-647.

[Cho et al. 2001] Cho, S., Gordon, P., Moore, J.E., Richardson, H.W., Shinozuka, M., Chang, S., Integrating Transportation Network and Regional Economic Models to Estimate the Cost of a Large Urban Earthquake, *Journal of Regional Science* V. 41(1), p. 39-65.

[Clarke et al. 1985] Clarke, S.H., Greene, H.G., Kennedy, M.P., Identifying Potentially Active Faults and Unstable Slopes Offshore in *Evaluating Earthquake Hazards in the Los Angeles Region - an Earth Science Perspective* USGS Professional Paper 1360, p. 347 - 373.

[Dolan et al. 1995] Dolan, J.F., Sieh, K., Rockwell, T.K., Yeats, R.S., Shaw, J., Suppe, J., Huftile, G., and Gath, E. (1995). Prospects for larger or more frequent earthquakes in greater metropolitan Los Angeles, California. *Science*, V.267, p. 199-205.

[Ellsworth 1990] Ellsworth, W. L. (1990). Earthquake history, 1769-1989, in Wallace, R. E. (ed.), *The San Andreas fault system, California. U. S. Geol. Surv. Professional Paper* 1515 p. 153-187.

[Gordon, Moore II, and Richardson 2002] Gordon, Moore II, and Richardson (2002) Economic-engineering integrated models for earthquakes: Socioeconomic impacts, Final Report to the Pacific Earthquake Engineering Research (PEER) Center, University of California, Berkeley, 1301 S. 46th St., Richmond, CA 94804-4698.

[Gordon et al. 1998] Gordon P., Moore II, J.E., Richardson, H. W., and Shinozuka, M., (1998). The socio-economic impacts of "the big one," Southern California Studies Center (SC^2), University of Southern California.

[Gordon, Richardson, and Davis 1998] Gordon, P., Richardson, H. W., and Davis, B., (1998). Transport-related impacts of the Northridge Earthquake, *Journal of Transportation and Statistics*, V.1, p. 21-36.

[Gorsline 1996] Gorsline, D., Depositional events in Santa Monica Basin, California Borderland, over the past five centuries. *Sedimentary Geology*, V. 104, p. 73 - 88.

[Greene et al. 2000] Greene, H.G., Maher, N., Paul, C.K., (2000). Landslide Hazards off of Santa Barbara, California (abstract). *American Geophysical Union Fall Meeting, 2000*.

[Gutenberg et al. 1932] Gutenberg, B., Richter, C., and Wood, H., (1932). The Earthquake in Santa Monica Bay, California on August 30, 1930. *Bulletin of the Seismological Society of America*, p. 138- 154.

[Hamilton et al. 1969] Hamilton, R.,R. Yerkes, R. Brown Jr. R. Burford, and J., DeNoyer (1969). Seismicity and associated effects, Santa Barbara Region in *Geology, Petroleum Development, and Seismicity of the Santa Barbara Channel Region, California. USGS Professional Paper 679-D*, p. 47 - 69.

[Hampton et al. 1996] Hampton, M.A., Lee, H.J., and Local, J. (1996). Submarine landslides. *Reviews of Geophysics*, V.34 (1), p. 33 – 59

[Hauksson 1990] Hauksson, E. (1990). Earthquakes, Faulting and Stress in the Los-Angeles Basin. *Journal of Geophysical Research - Solid Earth*, V.95 (BIO), p. 15365 - 15394.

[Hauksson and Gross 1991] Hauksson, E. and Gross, S. (1991). Source Parameters of the 1933 Long-Beach Earthquake. *Bulletin of the Seismological Society of America*, V.81 (1), p. 81 - 98.

[Hauksson and Saldivar 1986] Hauksson, E., and G. Saldivar (1986). The 1930 Santa Monica and the 1979 Malibu, California Earthquakes. *Bulletin of the Seismological Society of America*, V.76 (6), p. 1542 - 1559.

[Kawata et al. 1999] Kawata, Y., Benson, B., Borrero, J.C., Borrero, J.L., Davies, H.L., deLange, W.P., Imamura, F., Letz, H., Nott, J., and Synolakis, C.E. (1999), Tsunami in Papua New Guinea was as Intense as First Thought *Eos, Transactions of the American Geophysical Union*, V. 80 (9), p. 101, 104 - 105.

[Lander and Lockridge, 1989] Lander, J.F. and Lockridge, P.A., United States Tsunamis (including United States possessions) 1690 - 1988, *National Geophysical Data Center, Publication 41-2*, U.S. Department of Commerce, National Oceanic and Atmospheric Administration.

[Lander et al. 1993] Lander, J., P. Lockridge, and M. Kozuch (1993). *Tsunamis Affecting the West Coast of the United States*. U.S. Department of Commerce, National Oceanic and Atmospheric Administration.

[Locat et al. 2003] Locat, J., Locat, P., and Lee, H.J. (2003). Numerical analysis of the mobility of the Palos Verdes debris avalanche, California, and its implication for the generation of tsunamis. *Marine Geology*, in press. Available from the authors.

[Matsuyama et al. 1999] Matsuyama, M., Walsh, J.P., and Yeh, H. (1999). The effect of bathymetry on tsunami characteristics at Sisano Lagoon, Papua New Guinea. *Geophysical Research Letters*, V.26 (23), p. 3513-3516.

[McCarthy 1993] McCarthy, R.J., Bernard, E.N., Legg, M.R. (1993). The Cape Mendocino earthquake: A local tsunami wakeup call? In *Proceedings of the Eighth Symposium on Coastal and Ocean Management*, New Orleans, Louisiana, p. 2812 – 2828.

[McCulloch 1985] McCulloch, D., (1985). Evaluating Tsunami Potential in Evaluating Earthquake Hazards in the Los Angeles Region: *An Earth Science Perspective*. USGS Prof. Paper 1360, p. 375 - 414.

[Murty 1979] Murty, T.S. (1979). Submarine Slide-Generated Water Waves in Kitimat Inlet, British Columbia, Journal of Geophysical Research, V. 84, (C12), p. 7777-7779.

[Okal et al. 2002a] Okal, E.A., Fryer, G.J., Borrero, J.C. and Ruscher, C. (2001). The landslide and local tsunami of 13 September 1999 on Fatu Hiva (Marquesas Islands; French Polynesia). *Bulletin de la Societe Geologique de Prance* in press. Available from the authors.

[Richardson et al. 1993] Richardson, H.W., Gordon, P., Jun, M.J., and Kim, M. H., (1993). PRIDE and prejudice: The economic impacts of growth controls in Pasadena, *Environment and Planning A*, V.25A, p. 987-1002.

[Satake and Somerville 1992] Satake, K. and P. Somerville (1992). Location and size of the 1927 Lompoc, California Earthquake from tsunami data. *Bulletin of the Seismological Society of America*, V.82, p. 1710 - 1725.

[Stevens 1996] Stevens, B.H. (1996). PC I-O Version 7. Heightstown, New Jersey: Regional Science Research Corporation.

[Stevens, Treyz, and Lahr 1983] Stevens, B. H., Treyz G., and Lahr, M., (1983). A new technique for the construction of non-survey regional input-output models, *International Regional Science Review,* V.8, p. 271-286.

[Stierman and Ellsworth 1976] Stierman, D. and W. Ellsworth (1976). Aftershocks of the February 21, 1973 Point Mugu, California earthquake. *Bulletin of the Seismological Society of America*, V.66, (6), p. 1931 - 1952.

[Synolakis et al. 1997b] Synolakis C.E., McCarthy, R., Bernard, E.N., (1997). Evaluating the Tsunami Risk in California. In *Proceedings of the Conference of the American Society of Civil Engineers, California and the World Ocean '97*, San Diego, CA, p. 1225 - 1236.

[Synolakis et al. 2002] Synolakis, C.E., Bardet, J.P., Borrero, J.C., Davies, H.,0kal, E.A., Silver, E.A., Sweet, S., Tappin, D.R. (2002). Slump Origin of the 1998 Papua New Guinea Tsunami *Proceedings of the Royal Society, London*. In press, manuscript accepted in final form. Available from the authors.

[Tappin et al. 1999] Tappin, D.R., and 18 co-authors (1999). Sediment slump likely caused 1998 Papua New Guinea Tsunami. Eos, *Transactions of the American Geophysical Union*, V.80 (30).

[Tappin et al. 2001] Tappin, D.R., Watts, P., McMurty, G.M., Lafoy, Y., Matsumoto, T. (2001). The Sissano, Papua New Guinea tsunami of July 1998 - offshore evidence on the source mechanism. *Marine Geology*, V. 175, p. 1-23.

[Toppozada et al. 1981] Toppozada, T.,C. Real, and D. Parke (1981). Preparation of isoseismic maps and summaries of reported effects for pre — 1900 California Earthquakes. *California Division of Mines and Geology, Annual Technical Report. #81 - 11SA.* p. 34, 136 - 140.

[Watts 1998] Watts, P. (1998). Wavemaker curves for tsunamis generated by underwater landslides. *Journal of Waterway, Port, Ocean and Coastal Engineering*, V.124 (3), p. 127 - 137.

[Watts 2000] Watts, P. (2000). Tsunami features of solid block underwater landslides. *Journal of Waterway, Port, Ocean and Coastal Engineering*, v. 126 (3), p. 144 – 152.

[Watts et al. 2001] Watts, P., Gardner, J. V., Yalciner, A.C., Imamura, F. and Synolakis, C.E. (2001) Landslide Tsunami Scenario off Palos Verdes, California, *Natural Hazards*, In Press. Available from the authors.

THE EMERGING ROLE OF REMOTE SENSING TECHNOLOGY IN EMERGENCY MANAGEMENT

Beverley J. Adams, Ph.D.
Charles K. Huyck
ImageCat Inc. 400 Oceangate, Suite 1050, Long Beach, 90802

ABSTRACT

The use of remote sensing technology is becoming increasingly widespread at all levels of the disaster management cycle. This paper reviews a number of emerging application areas, drawing on examples from both natural disasters such as earthquakes, and man-made events including recent terrorist attacks. In terms of mitigation and preparedness, remote sensing is making an increasing contribution towards inventory development, hazard and risk assessment and logistical planning. After a disaster has occurred, remote sensing offers a rapid low-risk method of assessing damage to critical infrastructure and the urban built environment. In particular, high resolution coverage enabled the detection of bridge damage following the Northridge earthquake and collapsed buildings following various earthquakes in the U.S., Europe and Japan. Once the initial chaos has subsided, the value of remote sensing data lies in post-event monitoring; tracking changes in hazard location and extent. Reference is made to emergency operations following the World Trade Center attack, where remote sensing imagery supported various response efforts. In terms of secondary applications, details are given concerning innovative new research, using information derived from remote sensing coverage for running traffic rerouting programs, and as a supplementary data source for loss estimation models.

1   INTRODUCTION

The US Federal Emergency Management Agency (FEMA) defines an emergency as *'any natural or man-caused situation that results in or may result in substantial injury or harm to the population or substantial damage to or loss of property'* (FEMA, 2003b). These so called 'situations' include naturally occurring earthquakes, fires, floods, landslides and volcanic eruptions, together with man-made events such as terrorist attacks, oil spills, and recently, loss of the Columbia space shuttle.

Remote sensing technology has a central role to play in driving down the economic costs and reducing the human losses caused by extreme events. Its value as a supplement to existing emergency management strategies is increasingly recognized in national and international arenas (see Jayaraman *et al.*, 1997; also Kerle and Oppenheimer, 2002). In the US, a Policy Resolution drafted by the Western Governors Association in 2000 formally acknowledged the importance of Remote Sensing and Geographic Information System (GIS) technology in disaster situations (Western Governors Association, 2000). Today, numerous federal departments and agencies tasked with evaluating disaster information, rely on remote sensing coverage (for details, see Laben, 2002). Internationally, initiatives have been launched by organizations including the United Nations (UN), European Commission (EC), Commission for Earth Observation Studies (CEOS), and various space agencies. Currently, research efforts are being spearheaded through the UN International Decade on Natural Disaster Reduction (IDNDR, 2003) and its successor the International Strategy for Disaster Reduction (ISDR, 2003), the EC Natural Hazard Project (JRC, 2003), and the CEOS International Global Observation Strategy (IGOS, 2001). In terms of application to real emergency situations, the 2000 Charter on Space and Major Disasters was set up in the framework of the 1999 UNISPACE III conference of the United Nations (for a current status report, see International Charter, 2003). It represents an important commitment to extend the use of space facilities to aid emergency management, through promoting international cooperation (CEOS, 2001). The European

Space Agency now offers an Earth Watching Service, providing satellite disaster coverage acquired by Landsat, ERS and JERS sensors (ESA, 2003). Close to home, the FEMA website posts remote sensing coverage of recent events within the US (Mileti, 1999; FEMA, 2003a).

As a disaster management tool, remote sensing technologies offer significant contributions throughout the Emergency Management Cycle (see, for example, Walter, 1990, 1991; Luscombe and Hassan, 1993; Jayaraman *et al.*, 1997; also Williamson *et al.*, 2002). As shown by the conceptual representation in Figure 1, emergencies may be structured into phases of mitigation, preparedness, response and recovery (see also Garshnek and Burkle, 2000). Mitigation and preparedness measures serve to minimize the impact of future disasters, while response and recovery apply once an event has occurred. These stages are cyclic, with one feeding into the next.

Figure 1. Conceptual representation of the emergency management cycle. Definitions are drawn from the California Governors Office of Emergency Services (OES) document concerning Emergency Management (Davis and Jones, 2003)

While studies and reports concerning the application of remote sensing technology in disaster situations date back several decades (see, for example, Garafalo and Wobber, 1974; Richards, 1982; also Walter, 1990, 1991), this paper reviews recent research developments and emerging applications within the realm of emergency management. A cursory review of the literature indicates that this is an extremely broad topic. Consequently, the scope of the present study is narrowed to consider state-of-the-art research being conducted in several distinct areas of the disaster management cycle (see Figure 1). In terms of mitigation and preparedness, Section 2 considers the use of remote sensing for inventory development, hazard and risk assessment and logistical support. Section 3 goes on to explore its value during response and recovery phases. The initial detection of damage to infrastructure and built structures is described, drawing on examples from recent earthquakes and terrorist attacks. This is followed by brief reviews of remote sensing technology for post-event monitoring, and its integration into secondary models yielding important supplementary information, such as loss estimates. Following a summary of key findings in Section 4, future directions for research are considered in Section 5.

In all cases, documented techniques have attained 'proof of concept', or implementation status. For a comprehensive evaluation of the contributions made by remote sensing in the full range of emergency situations, readers are also referred to the Natural Disaster Reference Database of reports managed by NASA (NDRD, 2003), the overview of emerging US technologies by Mileti (1999), European pilot projects for natural hazard monitoring (see San Miguel-Ayanz et al., 2000), and the final report of the CEOS disaster management support group (CEOS, 2001).

2   MITIGATION and PREPAREDNESS

While working towards the long-term goal of disaster prevention, in the shorter term, contemporary emergency management is concerned with minimizing the extent and effects of extreme events (Garshnek and Burkle, 2000). Returning to the emergency management cycle in Figure 1, mitigation and preparedness activities are undertaken prior to natural and man-made disasters. Mitigation measures serve to reduce or negate the impact of an event, while preparedness efforts facilitate a more effective response once the disaster has occurred. The importance of integrating remote sensing data into pre-event emergency management strategies is increasingly acknowledged. The following sections present a number of illustrative examples (see also Walter, 1991), including ongoing research into the development and update of critical inventories, together with brief reviews of hazard and risk assessment, and improved logistical support.

2.1   Inventory development

Compiling a comprehensive and accurate database of existing critical infrastructure is a priority in emergency management, since it provides a basis for simulating probable effects through scenario testing, while setting a baseline for determining the actual extent of damage and associated losses once an event has occurred. In the context of mitigation and preparedness, demand is increasing for accurate inventories of the built environment, in order to perform vulnerability assessments, estimate losses in terms of repair costs (RMSI, 2003), assess insurers liability, and for relief planning purposes (Sinha and Goyal, 2001; RMSI, 2003). In lesser developed regions of the world, inventories are often scarce. CEOS (2001) documents a program to compile comprehensive records of urban settlements that could be affected in the event of an earthquake, to avoid a repeat of the 1998 Afghanistan earthquake, when due to unavailability of even simple maps or images, relief workers experienced extreme difficulty locating affected villages.

Although the location of urban centers is generally well documented for developed nations, interest is growing in accurate low-cost methods for characterizing the built environment in more detail. Building inventories are a primary input to loss estimation models, such as the FEMA program HAZUS (Hazards-US) and California State system EPEDAT (Early Post-Earthquake Damage Assessment Tool). These are used as planning tools prior to an event, and as a response tool once an event has occurred. Measures of interest include: building height; square footage; and occupancy (use). To a large degree, the accuracy of loss estimates depends on the quality of input data. Default datasets are often based on regional trends, rather than local specifics. Research being undertaken by ImageCat, in conjunction with the Multidisciplinary Center for Earthquake Engineering Research (MCEER), suggests that remote sensing data offers a detailed inventory of both height and square footage, which through supplementing existing datasets, may lead to more accurate loss estimates.

Figure 2 conceptualizes the methodological procedure through which building height and square footage can be obtained from a combination of interferometric SAR (IfSAR) and optical imagery (see also Eguchi *et al.*, 1999). Figure 2a shows the derivation of buildings heights, in terms of a normalized digital surface model (nDSM). Based on the method developed by Huyck *et al.* (2002), this nDEM is obtained as the difference between a SAR-derived digital surface model (DSM) and a bare-earth digital terrain model (DTM). The former DSM represents the apparent ground surface, as a composite of superimposed features, such as buildings and underlying bare earth topography. The latter DTM is solely topographic, obtained from the same base data via a sequence of filters. As shown by the flowchart in Figure 2b, building heights are recorded as the local maxima within footprints delineated on high-resolution aerial photography. While the present set of building outlines were defined manually, a review of the literature suggests that research is progressing towards automated extraction procedures (see, for example, Collins *et al.*, 1995; Heuel and Kolbe, 2001; Noronha and Nevatia, 2001; Nevatia and Price, 2002). The heights are then translated to stories, using a conversion factor that corresponds with standard loss estimation software (see HAZUS, 1997). Ground level square footage is also recorded on a per building basis, as the footprint area in pixel units. Using a scaling factor based on image resolution, this value is converted to single storey square footage. Finally, the total square footage for each structure is computed as the product of the number of stories and ground level area.

The efficacy of this methodology has been tested for case study areas in Los Angeles, where the values for building height and coverage correspond closely with independently derived tax assessor data (Eguchi *et al.*, in preparation). Moving forwards, methodological procedures are under development that use these results to update existing inventories within the HAZUS program (for an extended discussion, see Section *3.3*).

A significant advantage of remotely-derived inventories is the relative ease with which they can be updated. This is particularly important at a city wide scale, where the overview offered by satellite imagery can be used by planning departments to track urban growth (see DOT/NASA, 2002, 2003). Classifying imagery into vegetation, concrete, and buildings is a straight-forward task, which is readily applied to multi-temporal coverage. Growth is detected in terms of change between the scenes.

In addition to building-related catalogues, remote sensing is being used to map other critical infrastructure. For example, high resolution imagery provides a detailed overview of transportation networks from which street centerlines can be extracted (for a useful review, see Xiong, 2003). The resulting data are widely used within digital routing programs and hard-copy mapping applications.

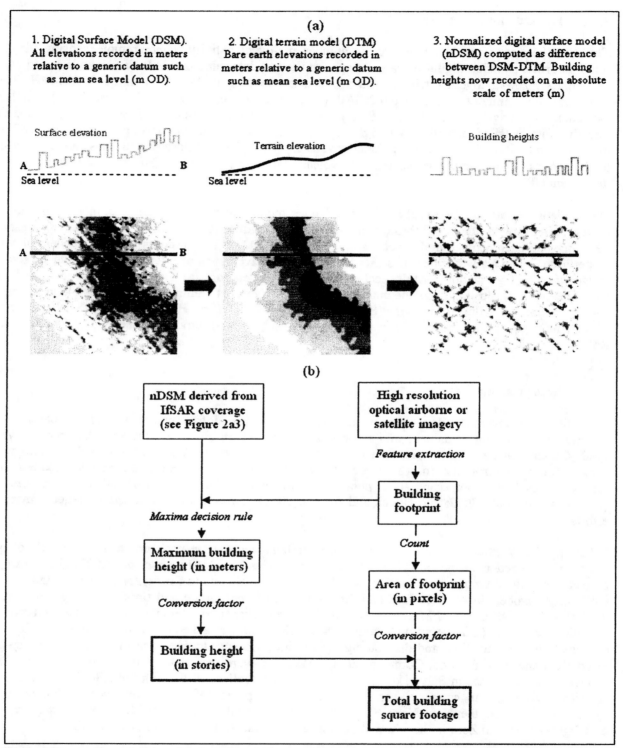

Figure 2 Conceptual representation of methodological procedures used to obtain building inventory data from remote sensing coverage. (a) The derivation of a normalized digital surface model (nDSM) from IfSAR data, as a basis for building height measurements. (b) Processing steps involved in computing building height (in stories) and coverage (in total square footage), in a format suitable for supplementing default data in loss estimation programs such as HAZUS.

2.2   Hazard and risk assessment

The identification of probable hazards is another pre-event activity where remote sensing plays an important role. Detailed digital elevation maps are used to determine probable dam inundation in MIKE 21 (Mike21, 2002), together with fluvial and coastal flooding in MIKE 21 and HAZUS-MH. These elevations are routinely derived from IfSAR (see, for example, Galy and Sanders, 2000) and Light Detection and Ranging (LIDAR) data. Risk maps showing landslide potential, use remotely sensed data directly from detailed elevation readings such as these, and indirectly through geological, soil and moisture information from optical and radar coverage (CEOS, 2001). Interferometry has also been used to track changes in topography associated with volcanic activity (JPL, 1995; Lu et al., 2003), and glacial movement (JPL, 2003).

Optical data are particularly useful for the visual assessment of hazards. For example, monitoring patterns of vegetation growth, identified through classification techniques (Campbell, 1996), provides a means of detecting encroachment around pipelines (DOT/NASA, 2003). This process is most successful when 'supervised' by an analyst, whereby a user identifies indicative 'areas of interest' to guide the subsequent image-wide categorization. Multispectral coverage extending to longer wavelengths of the electromagnetic spectrum, offers the unique opportunity to inspect features that are invisible to the naked eye. In terms of wildfire risk, the Southern California Wildfire Hazard Center (SCWHC, 2003) documents the quantification of chaparral fuel content using multispectral data (see also CEOS, 2001; and Roberts et al., 1998).

2.3   Logistical support

In addition to inventory development, databases of critical infrastructure provide a baseline for determining the extent of damage and associated losses once an event has occurred. Taking the World Trade Center attack as an example, the successful role played by remote sensing and GIS technologies in response efforts at Ground Zero, was largely attributable to the prior existence of a very detailed base map for New York City. Comprising aerial photos and GIS data for building footprints, roads, and lifelines (see Cahan and Ball, 2002; and Huyck and Adams, 2002), these data underpinned subsequent mapping efforts.

Following the 9/11 terrorist attack, it was widely recognized that several remote sensing technologies were underutilized in response efforts (Huyck and Adams, 2002; Huyck et al., 2003). For example, correctly calibrated temperature readings would have been valuable to firefighters, but were unavailable until early October. In order to facilitate the collection of appropriate and timely coverage of extreme events within the U.S., FEMA and NASA have established a Remote Sensing Consultation & Coordination Team (Langhelm and Davis, 2002). This team is tasked with identifying suitable data, coordinating its acquisition, and distributing the resulting imagery to everyone involved in the response effort (R. Langhelm, Personal Communication). Members were in a state of readiness during the 2002 Winter Olympic Games in Salt Lake City. To support data collection through the RSCCT system, there are clear advantages in having contractual agreements in place before an event occurs. Indeed, prior agreements between the New York State Office for Technology and EarthData paved the way for overflights of Ground Zero in the aftermath of the terrorist attack (Huyck and Adams 2002).

3   RESPONSE and RECOVERY

Response and recovery phases of the emergency management cycle refer to relief efforts once an event has occurred. In terms of response (for the California OES definition, see Figure 1), remote sensing technology offers a number of distinct advantages over traditional ground-based techniques, through the provision of alternative routes for critical information flows (see Puzachenko *et al.*, 1990; Garshnek and Burkle, 2000). Following the onset of an extreme event, assessing the nature, extent, and degree of damage are priorities. However, these tasks are often problematic, due to the distributed nature of natural disasters, and limited accessibility when transportation routes are disrupted. Accordingly, Section 3.1 outlines methodological procedures for the remote detection of damage to urban settlements and critical highway infrastructure. After the initial chaos has subsided, emergency efforts turn to monitoring activities and the provision of logistical support. Section 3.2 describes the role of remote sensing technology during these post-event operations, featuring brief examples from a range of natural and man-made disasters. In emergency situations, remote sensing data is clearly an important source of information in its own right. However, in some situations, the data yielded through analytical techniques has additional value, as the input to secondary models. In recognition, Section 3.3 reviews several novel approaches to loss estimation, which employ derived inventory and damage data.

3.1   Damage detection

Real time damage detection following a natural or man-made disaster initiates the response process, providing the information needed to: (a) prioritize relief efforts (Lavakare, 2001); (b) direct first responders to critical locations, thereby optimizing response times (Sinha and Goyal, 2001) and ultimately saving lives; (c) compute initial loss estimates (see Section 2.1 and Section 3.4, also RMSI, 2003; and Tralli, 2000); and (d) determine whether the situation warrants national or international aid. Of particular importance is damage sustained by urban settlements, together with critical infrastructure, such as roads, pipelines and bridges. For the present study, damage detection methodologies are described for highway bridges and buildings, drawing on research conducted following recent earthquake events and experience gained in the aftermath of the World Trade Center attack. The methodological process follows either a *direct* and *indirect* approach. In the former case, damage is detected by directly observing the characteristics of, or temporal changes to an object of interest. In the latter case, damage is detected through a surrogate indicator.

In extreme events, such as natural disasters and terrorist attacks, the performance of critical transportation elements is a major concern. Taking the U.S. as an example, the transportation network is vast, comprising over 500,000 bridges and 4 million miles of road (Williamson *et al.*, 2002). When a disaster like the 1994 Northridge earthquake strikes, effective incident response demands a rapid overview of damage sustained by numerous elements, spread over a wide geographic area. Given the magnitude and complexity of transportation systems, near-real time field-based assessment is simply not an option. Taking the recent earthquake centered on the Bingol province of Turkey as another example, the media reported damage to roads and bridges, with a number of villages cut off (ABC, 2003). Considering the critical 48 hour period that urban search and rescue teams have to locate survivors, accessibility must be quickly and accurately determined, in order to reroute response teams and avoid life threatening delays. Irrespective of whether the event occurred in Turkey or the US, earth orbiting remote sensing devices like IKONOS and Quickbird present a high-resolution, synoptic overview of the highway system, which can be used to monitor structural integrity and rapidly assess the degree of damage.

Under the auspices of a DOT/NASA initiative promoting remote sensing applications for transportation (Morain, 2001; DOT/NASA, 2002, 2003), preliminary damage detection algorithms termed *'Bridge Hunter'* and *'Bridge Doctor'* have been developed for highway bridges (Adams *et al.*, 2002). From the methodological summary in Figure 3, Phase 1 of the damage detection process employs Bridge Hunter to track down and compile a catalogue of remote sensing imagery, together with attribute information from Federal Highway Administration Databases (FHWA) databases. During Phase 2, Bridge Doctor diagnoses the 'health' of bridges, determining whether catastrophic damage has been sustained. In this case, the bridge damage state is quantified directly, in terms of the magnitude of change between a temporal sequence of images acquired 'before' (Time 1) and 'after' the event (Time 2). It is hypothesized that for collapsed bridges, where part of the deck fell or was displaced, substantial changes will be evident on the remote sensing coverage. However, where negligible damage was sustained, change should be minimal.

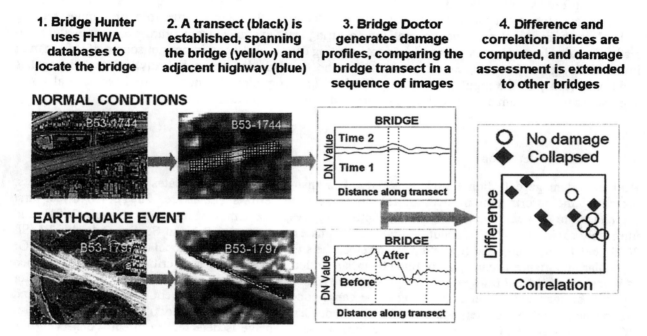

Figure 3 Schematic summary of the 'Bridge Hunter' and 'Bridge Doctor' damage detection methodologies, with examples of results obtained for collapsed versus non-damage bridges following the 1994 Northridge earthquake.

The Northridge earthquake was employed as a test bed for model development. Widespread damage was sustained by the transportation network when the 6.7 magnitude event struck Los Angeles on the 17th January 1994. Six examples of bridge collapse were available for model calibration and validation. Damage profiles obtained from SPOT imagery clearly distinguish between these extreme scenarios. From the subset of results in Figure 3, reflectance signatures for the non-damaged example are consistent at Time 1 and Time 2, following a similar pattern along the highway and across the bridge. For the collapsed scenario, substantial changes are evident between the 'before' and 'after' earthquake scenes. The damage profiles no longer follow a similar trend, with abrupt divergence in signature around the collapsed span. Damage indices including difference and correlation offer a quantitative comparison. The bivariate damage plot clearly distinguishes between the low correlation and high difference associated with collapsed bridges, and high correlation and low difference of their non-damaged counterparts.

This damage detection methodology is now being tested in Salt Lake City, Utah, where very high resolution coverage is available for a number of highway segments and bridges that underwent reconstruction prior to the 2000 Winter Olympics. The temporal changes due to rebuilding are analogous to those observed following an earthquake event.

The use of remotely sensed data for assessing building damage offers significant advantages over ground-based survey. Where the affected area is extensive and access limited, it presents a low-risk, rapid overview of an extended geographic area. A range of assessment techniques are documented in the literature, including both direct and indirect approaches. In the former case, building damage is recorded directly, through its signature within the imagery (for a useful review, see Yamazaki, 2001). Research by Matsuoka and Yamazaki (1998), Chiroiu et al. (2002) and Chiroiu and Andre (2001) suggests that collapsed and extensively damaged buildings have distinct spectral signatures. However, moderate and minor damage states are indistinguishable from non-damage. Damage is usually quantified in terms of the extent or density of collapsed structures. In the latter case, damage may also be determined using an indirect indicator, based on the theory that urban nighttime lighting levels diminish in proportion to urban damage (CEOS, 2001). Further details of the respective methodologies are given below.

Direct approaches to building damage assessment may be categorized as multi- and mono-temporal. Following a similar theoretical basis to the bridge damage methodology described in Section 3.1.1, *multi-temporal* analysis determines the extent of damage from spectral changes between images acquired at several time intervals; typically before and after an extreme event. Figure 4 outlines the methodological process that has been employed at city-wide and regional scales for various earthquakes, using optical and Synthetic Aperture Radar (SAR) imagery.

At a city-wide scale, comparative analysis of Landsat and ERS imagery collected before and after the 1995 Hyogoken-Nanbu (Kobe) earthquake, suggested a trend between spectral change and ground truth estimates for the concentration of collapsed buildings (Aoki et al., 1998; Matsuoka and Yamazaki, 1998, 2000a, 2000b; Tralli, 2000; Yamazaki, 2001). Similar qualitative and quantitative methods were used to evaluate damage in various cities affected by the 1999 Marmara earthquake in Turkey (Eguchi et al., 2000a, 2000b). Visual comparison between SPOT scenes in Figure 5a-b for the town of Golcuk, demonstrates changes in reflectance due to earthquake damage (see also Estrada et al., 2001a, 2001b). Areas of pronounced change are highlighted by circles. Figure 5c-f shows measures of change such as difference, correlation and block correlation (see also Eguchi et al., 2003), overlaid with the zones where ground truth data were collected (AIJ, 1999). Graphing the concentration of building damage by each measure generates the damage profiles in Figure 6 (see also EDM, 2000; Huyck et al., 2002; Eguchi et al., 2002, 2003). There is a clear tendency towards increased offset between before and after scenes as the percentage of collapsed structure rises from class A-E.

This methodology has also been implemented for ERS SAR coverage (Eguchi et al., 2000b), which offers advantageous 24/7, all weather viewing, and an additional index of change termed coherence (see also Matsuoka and Yamazaki, 2000a; Yamazaki, 2001; Huyck et al., 2002; and Eguchi et al. 2003). Matsuoka and Yamazaki (2002, 2003) have recently generalized this approach, to show consistency in the trend between building collapse and remote sensing measures for the 1993 Hokkaido, 1995 Kobe, 1999 Turkey, and 2001 Gujurat earthquakes.

At a regional scale, Matsouka and Yamazaki (2002) detect damaged settlements within Marmara and Gujurat provinces, following 1999 and 2001 earthquakes in Turkey and India. This approach provides a quick-look assessment of the damage extent, and directs responders to the severely hit areas. For further details of multi-temporal damage detection following the Gujurat event, readers are also referred to Yusuf et al. (2001a, 2000b, 2002), Chiroiu et al. (2002, 2003) and Chiroiu and Andre (2001). For the 2001 El Salvador earthquake, see Estrada et al. (2001a).

Mono-temporal analysis detects damage from imagery acquired after a disaster has occurred. It is particularly useful where 'before' data is unavailable. The methodology relies on direct recognition of collapsed structures on high-resolution coverage, through either visual recognition or diagnostic measures. As with the multi-temporal approach, it is most effective for extreme damage states, where buildings have collapsed or are severely damaged (Chiroiu et al., 2002).

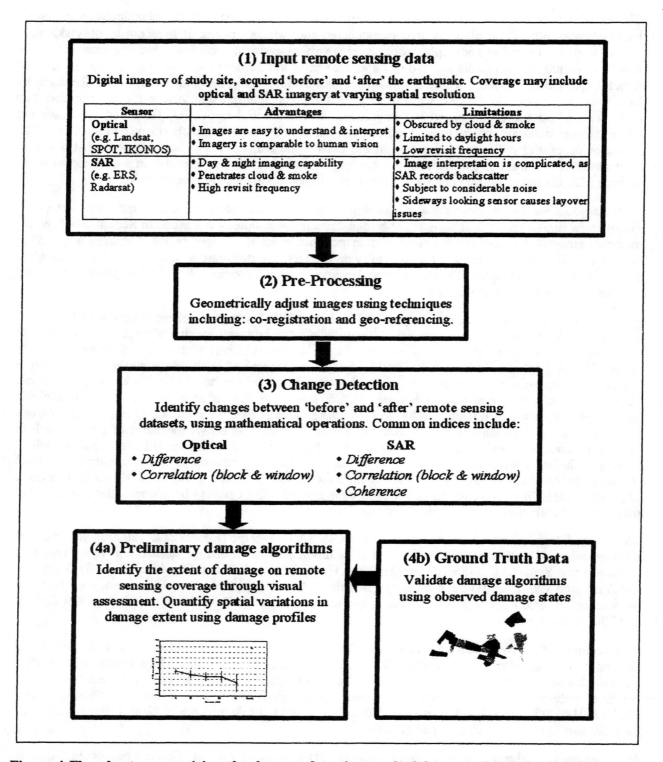

Figure 4 Flowchart summarizing the damage detection methodology employed for buildings and urban settlements, using multi-temporal remote sensing imagery.

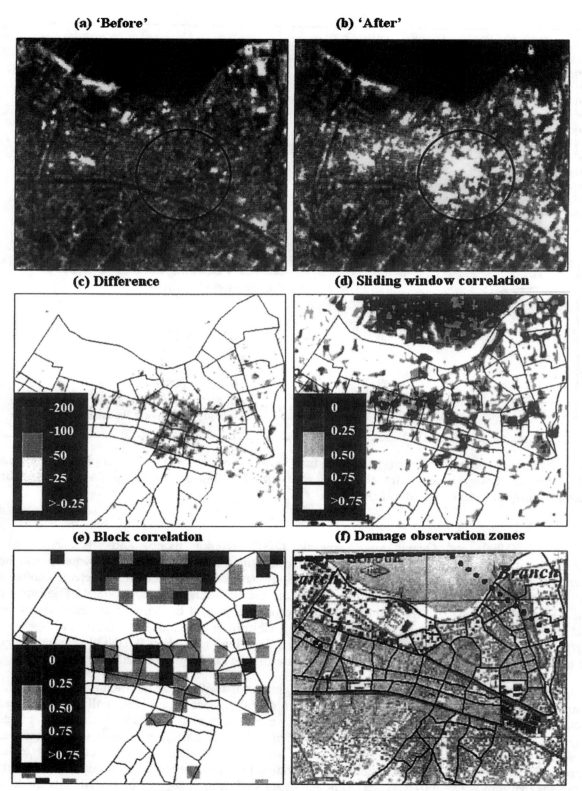

Figure 5 Panchromatic SPOT4 coverage of Golcuk, showing (a) 'before' image; (b) 'after' image; (c) difference values; (d) sliding window correlation; (e) block correlation; and (f) ground truth zones, where the percentage of collapsed buildings was observed. Data courtesy of ESA, NIK and AIJ.

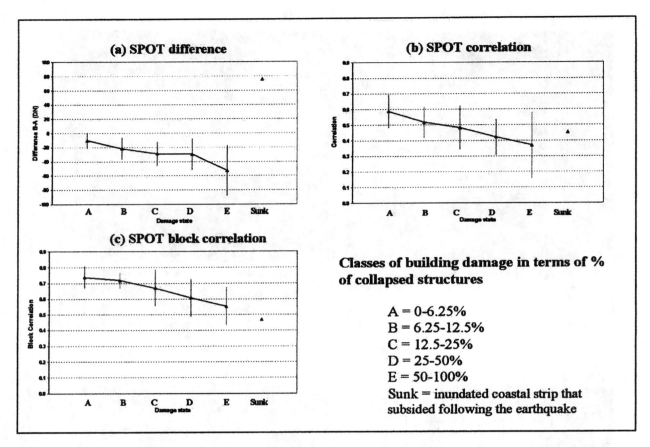

Figure 6 Damage profiles for Golcuk, showing how values recorded in the 70 sample zones for each SPOT index of change varies with the concentration of collapsed buildings (A-E). Error bars represent 1 standard deviation about the mean

Ogawa *et al.* (1999) and Ogawa and Yamazaki (2000) employ mono- and stereoscopic photo interpretation of vertical aerial photography to determine the damage sustained by wooden and non-wooden structures in Kobe. A 'standard of interpretation' was devised to distinguish between collapsed, partially collapsed, and non-damage structures, based on: the occurrence of debris; level of deformation; and degree of tilt. Success of this methodological approach is judged in terms of correspondence with ground truth observations. Chiroiu and Andree (2001), Chiroiu *et al.* (2002) and Saito and Spence (in review) use similar criteria to interpret building damage from high resolution IKONOS satellite imagery of the city of Bhuj, which sustained extensive damage during the 2001 Gujurat earthquake.

High speed automated aerial television is also emerging as a useful tool for mono-temporal damage assessment. Ogawa *et al.* (1999) and Hasegawa *et al.* (2000) inventory building collapse from visual inspection of HTTV imagery for Kobe. Diagnostic characteristics of debris and structural building damage are expressed quantitatively by Hasegawa *et al.* (1999) and Mitomi *et al.* (2002). Their basic methodology recognizes collapsed and non-damage scenarios in terms of color, edge and textural information. Multi-level slice and maximum likelihood classifiers determine the spatial distribution of these classes (Mitomi *et al.*, 2001b, 2002). Although developed using imagery of Kobe, this methodology has successfully detected collapsed buildings in Golcuk, Chi Chi (Mitomi *et al.*, 2000, 2001b) and Gujurat (Mitomi *et al.*, 2001a; also Yamazaki, 2001).

An *indirect* method of mono-temporal building damage assessment is also documented in the literature. In this instance, damage to building stock is inferred using a surrogate measure. Hashitera *et al.*, (1999) and Kohiyama *et al.* (2001) compare night-time lighting levels in US Defense Meteorological Satellite Program Operational Linescan System (DMSP-OLS) imagery acquired before and after the Marmara and Gujurat earthquakes. In both cases, areas exhibiting the greatest reduction in intensity corresponded with damaged settlements, supporting the hypothesis that fewer lights shine where buildings are severely damaged (Chiroiu and Andree, 2001). Operating under the cover of darkness, this damage assessment tool is a useful supplement to optically-based methodologies that are limited to daylight hours.

Although examples used to illustrate the preceding methodologies are drawn from earthquake events, damage detection from remote sensing imagery also proved particularly useful in the aftermath of the World Trade Center terrorist attack (Cahan and Ball, 2002; Hiatt, 2002; Huyck and Adams, 2002; Logan, 2002; Thomas *et al.*, 2002; Williamson and Baker, 2002; Huyck *et al.*, 2003). IKONOS coverage acquired on 12th September 2001 and posted on the Internet, provided people around The World with an early visualization of the devastation at Ground Zero. The first detailed pictures were captured the following day; the Fire Department of New York (FDNY) recorded oblique shots from a circling helicopter, and Keystone Aerial Surveys vertical photographs for the New York State Emergency Management Office. From the 15-16th September until mid October, EarthData systematically acquired orthophotographs, thermal and LIDAR data (for a full timeline of data acquisition, see Huyck and Adams, 2002). While these datasets were initially used to detect damage, in respect of their extended temporal coverage, further discussion of their usefulness is reserved for the following evaluation of the role played by remote sensing technology in protracted post-event monitoring.

3.2   Post event monitoring

As response efforts unfold following an extreme event, remote sensing is an important source of logistical support. The following section presents selected instances (see also, Mileti, 1999) where remote sensing has aided response efforts in the aftermath of man-made and natural disasters.

In terms of man-made disasters, remote sensing data was of value following the recent explosion of the Columbia Space Shuttle. A combination of airborne COMPASS and radar satellite imagery was used to show the distribution of the debris field (NOAA, 2003; Oberg, 2003). Returning to the 2001 World Trade Center attack (see also Section 3.1.2), LIDAR, thermal imagery and aerial photography acquired by EarthData gave a detailed overview of Ground Zero. Multitemporal analysis enabled the monitoring of cleanup operations. Volumetric analysis using LIDAR elevation data, such as that depicted in Figure 7, was used to track progress with clearing the debris pile. In several instances, the fusion of key datasets provided responders with valuable new information (Huyck and Adams, 2002; Huyck *et al.*, 2003). For example, overlaying the 3D LIDAR representation of the debris pile with a map of hazardous materials and fuel sources enabled firefighters to assess what was happening underneath the ground. The correlation between voids and the position of fuel and Freon tanks presented a focus for firefighting efforts, possibly preventing explosions that would have released toxic gases. The thermal data was also useful (see Rodarmel *et al.*, 2002). Overlaid with a two-dimensional 75x75ft transparent reference grid established by the FDNY, it provided a common system for tracking objects and remains amongst the debris. When fused with an orthophotograph, it facilitated strategic planning, which needed to consider the location of hotspots within the pile. It was also used to evaluate firefighting strategies, by visually noting differences in a time series of images during which various chemicals were tested. The aerial photographs were widely employed as a base-map. Applications included overlay with CAD models of floor plans for the Twin Towers, enabling search and rescue teams to pin point specific infrastructure, such as stairwells and elevator shafts.

Figure 7 Map showing a 3D terrain model for Ground Zero, produced from LIDAR data acquired by EarthData on September 19th 2001.

For natural disasters, remote sensing applications typically focus on tracking the location and extent of a given hazard, using a temporal sequence of images. In the case of wildfires, the online GEOMAC service (GEOMAC, 2003) integrates MODIS thermal imagery. Readers are referred to Ahern *et al.* (2001) for further details of the sensing process, and CEOS (2001) for a review of contemporary applications. While GEOMAC offers reasonably timely visualization at a regional scale, until a constellation of low earth orbiting satellites (LEOS) comes online (see, for example, Sun and Sweeting, 2001), the ultimate target of real time detection with 15 minute updates (CEOS, 2001) remains out of reach. For tracking floods, optical imagery has been widely used (see, for example, Sharma *et al.*, 1996; also Laben, 2002), despite the persist challenge posed by cloud cover. A number of authors illustrate all weather capability through integrating of optical and SAR imagery (Profeti and MacIntosh; 1997; and Tholey *et al.*, 1997; Wang *et al.*, 2003). Volcanic eruptions also represent a considerable challenge, creating a range of land- and air-based hazards. Kerle and Oppenheimer (2002) describe the use of optical and radar imagery to track fast flowing lahars. Monitoring the spread of atmospheric ash clouds is a further application area (see CEOS, 2001; also Francis and Rothery, 2000), which promises reduced risk to aviators.

In terms of man-made disasters, remote sensing is increasingly employed to track oil spills. For example, Danish and Norwegian agencies use satellite and airborne surveillance operationally, to perform reconnaissance on detected slicks. CEOS (2001) and Fingas and Brown (1997) note that optical, SAR and laser flurosensor devices are particularly useful for detecting and monitoring oil slicks. Tracking pollution and particulate debris is another emerging application. Atmospheric pollutants are recorded through increased absorption at specific wavelengths of the electromagnetic spectrum. Following the World Trade Center attack, hyperspectral imagery was recorded by the JPL Advanced Very High Resolution Imaging Spectrometrer (AVIRIS). Through studying absorption patterns in narrow bands, it was possible to map the concentration of airborne particulates surrounding Ground Zero, including concrete, cement and asbestos (Clark *et al.*, 2001). The Airborne LIDAR Pipeline Inspection System (ALPIS) uses an infrared laser to monitor gas plumes at ground surface level (LaSen, 2003). Together, these examples clearly point towards the potential application of spatial technology in response to bioterrorism, and the detection of airborne contaminants (see also Brown, 2002).

3.3 Supplying secondary models

In addition to the direct support of response efforts, data derived from remote sensing imagery has a secondary role to play through its integration into applied models. This section of the report discusses the extended application of inventory (see Section 2.1) and damage data (Section 3.1) within loss estimation and traffic routing programs.

The HAZUS program developed by FEMA (HAZUS, 1997) is a widely excepted framework for analyzing losses arising from earthquakes, which will soon be expanded to address hurricanes and flooding. The default building data delivered with HAZUS is based on generic assumptions made from census data and statistics provided by Dun and Bradstreet. For example, HAZUS assumes that *all* 'RES1' single family residential dwellings occupy 1,500 sq ft per unit. There are no automated adjustments to this figure by income or population density. Commercial structures classified as 'COM1', or retail trade, are ubiquitously assigned an area of 14,000 sq ft. Theoretically, users can update and modify HAZUS defaults using ground truth or tax assessor databases. Practically, these data are often unavailable, or the cost of obtaining data is unreasonable. It was shown in Section 2.1 that key measurements of building height and square footage can be generated from DEM elevation and optical data, facilitating the augmentation of US databases for HAZUS and other loss estimation programs.

It was also noted in Section 2.1 that within the international arena, inventories of this nature are scarce. However, using the techniques shown in Figure 2, it would be possible to generate a building stock estimate from square footage and height measures almost anywhere in the world. Global inventories of these measurements could be posted online, and in the event of an earthquake, applied in tandem with the building damage detection methodology (Section 3.1.2), to yield a rapid estimate of the damage extent and associated losses. Chiroiu and Andre (2001) and Chiroiu *et al.* (2002) also employ damage estimates for loss estimation, although in this instance, to quantify casualties arising from the Bhuj earthquake.

The detection of damage to highway infrastructure using remote sensing coverage is another application area that has increased potential through the use of secondary models. In this case, the location of collapsed bridges can be input to traffic routing models, such as the MCEER funded REDARS (Risk from Earthquake Damage to Roadway Systems) program, to compute alternative 'optimal' routes avoiding the obstruction (for details, see Werner *et al.*, 2003). Simulations have been run, based on the location of structures that collapsed in Los Angeles during the 1994 Northridge earthquake. It is straightforward to envisage how these results could be used to safely direct emergency crews around the obstruction. The REDARS program also has loss estimation capabilities, enabling the prediction of costs associated with traffic diversion while the bridge is replaced.

4   SUMMARY OF KEY FINDINGS

❖   Remote sensing technology is playing an increasing role at all levels of the emergency management cycle, where its use promises to drive down the economic costs, and further reduce the human losses caused by extreme events.

❖   Key application areas involve using optical and multispectral imagery, radar, IfSAR elevation and LIDAR data, acquired by satellite and airborne sensors prior to and in the aftermath of both natural disasters and man-made terrorist attacks.

❖   Imagery acquired 'before' an event enables a baseline case to be established. In terms of mitigation and preparedness, remote sensing data is particularly valuable for developing building and infrastructure inventories prior to earthquakes, and hazard and risk assessment for floods, brushfires and landslides. Measures such as data collection agreements and establishing advisory panels also provide further logistical support that should facilitate rapid response.

❖   Data collected 'after' an event offers a low-risk synoptic approach towards damage detection for critical infrastructure and urban environments. In terms of response and recovery, remote sensing data enables the rapid detection of damage to buildings and critical infrastructure such as highway bridges, together with post-event monitoring to support clean-up operations after events including the World Trade Center attack and Columbia disaster.

❖   In addition to direct response and recovery benefits, remote sensing imagery can be run through secondary models, to yield additional information. The input of building damage data to loss estimation software, such as HAZUS, supports national and international loss estimation. Feeding the location of damaged highway bridges through routing software such as the MCEER funded REDARS program, generates optimal routes for responders, together with an estimate of the losses caused by traffic diversion.

5   FUTURE DIRECTIONS

In order to effectively use remote sensing data and overcome potential stumbling blocks for its institutional acceptance, emergency management agencies must undertake further preparatory measures. At a fundamental level, the emergency management community requires further education about data that is available, together with its potential uses. In turn, the remote sensing industry needs to target specific application areas, necessitating improved understanding of emergency responders' data requirements in a full range of emergency situations. To promote the widespread use of remote sensing data alongside other spatial technologies, this must include education on a technical level. Essentially, the successful integration of remote sensing throughout the emergency management cycle will depend on familiarizing key personnel with this data before other events occur, so that they can become comfortable with it, and issues involved in its use. Through this education process, emergency management personnel will be able to communicate their needs to the data analysts.

Speed of data delivery is a key consideration that will further promote the use of remote sensing in emergency response. Taking the World Trade Center response efforts as an example, EarthData achieved an unprecedented turnaround time of 12 hours between the acquisition of aerial imagery and its delivery to responders. However, to maximize the value of this data in terms of currency, responders ideally require its delivery within a 3 hour time window (Huyck and Adams, 2002). Research effort is clearly required to accelerate download, processing and delivery timescales. Following delivery, the rapid production of derived information would be assisted by promoting widespread base data availability through file sharing, and standardization of the processing environment through the use of predetermined file structures and formats.

The timeliness of remote sensing datasets is a further area for improvement. At present, the high-resolution civilian systems, such as IKONOS and Quickbird, provide an excellent level of spatial detail. However, the frequency of cover is poor. Kerle and Oppenheimer (2002) identify future generations of LEOS as the way forward for remote sensing in emergency management. It may be envisaged that once in place, this system will provide continuous coverage in a manner similar to the operational global positioning system (GPS).

5   REFERENCES

ABC   (2003)   200   remain   trapped   after   Bingol   earthquake, http://www.abc.net.au/news/justin/nat/newsnat-1may2003-79.htm

Adams, B.J., Huyck, C., Mansouri, B., Eguchi, R. and Shinozuka, M. (2002) Post-disaster bridge damage assessment, *Proceedings of the 15th Pecora Conference: Integrating Remote Sensing at the Global, Regional, and Local Scale*, Denver.

Ahern, F., Goldammer, J.G. and Justice, C. (2001) *Global and Regional Vegetation Fire Monitoring from Space: Planning a Coordinated International Effort*, SPB: The Hague.

AIJ (1999) *Report on the Damage Investigation of the 1999 Kocaeli Earthquake in Turkey*, AIJ: Tokyo.

Aoki, H., Matsuoka, M. and Yamazaki, F., 1998, Characteristics of satellite SAR images in the damaged areas due to the Hyogoken-Nanbu earthquake, *Proceedings of the 1998 Asian Conference on Remote Sensing*, http://www.gisdevelopment.net/aars/acrs/1998/ts3/ts3007.shtml.

Brown, S.F. (2002) How infotech can combat homeland insecurity. Building America's anti-terror machine, *Fortune*, July 22: 99-104.

Cahan, B. and Ball, M. (2002) *GIS at Ground Zero: Spatial technology bolsters World Trade Center response and recovery*, http://www.geoplace.com/gw/2002/0201/0201wtc.asp

Campbell, J.B. (1996) *Introduction to Remote Sensing (2nd ed.)*, Taylor & Francis: London

CEOS (2001) *The Use of Earth Observing Satellites for Hazard Support: Assessments & Scenarios. Final Report of the CEOS Disaster Management Support Group*, http://www.oosa.unvienna.org/SAP/stdm/CEOS_DMSG_Final_Report.pdf

Chiroiu, L. and Andre, G. (2001) Damage assessment using high resolution satellite imagery: application to 2001 Bhuj, India earthquake, www.riskworld.com

Chiroiu, L., Andre, G., Guillande, R., and Bahoken, F. (2002) Earthquake damage assessment using high resolution satellite imagery, *Proceedings of the 7th U.S. National Conference on Earthquake Engineering*, Boston.

Chiroiu, L., Andre, G. and Bahoken, F. (2003) Earthquake loss estimation using high resolution satellite imagery, www.gisdevelopment.net/application/natural_hazards/ earthquakes/nheq0005.htm

Clark, R. N., Green, R.O., Swayze, G.A., Hoefen, T.M., Livo, K.E., Pavi, B., Sarcher, C., Boardman, J. and Vance, J.V. (2001) *Images of the World Trade Center Site Show Thermal Hot Spots on September 16 and 23, 2001*, Open File Report OF-01-405, U.S. Geological Survey

Collins, R.T., Hanson, A.R., Riseman, E.M. and Schultz, H (1995) Automatic extraction of buildings and terrain from aerial images, *International Workshop on Automatic Extraction of Man-Made Objects from Aerial and Space Images. Ascona, Switzerland,* 169-178.

Davis, G. and Jones, D. (2003) Emergency Management in California, Governors Office of Emergency Services, http://www.oes.ca.gov/Operational/OESHome.nsf/PDF/EMGuide/$file/EMGuide.pdf

DOT/NASA (2002) *Achievements of the DOT-NASA Joint Program on Remote Sensing and Spatial Information Technologies. Application to Multimodal Transportation,* http://www.ncgia.ucsb.edu/ncrst/synthesis/

DOT/NASA (2003) *Remote Sensing and Spatial Information Technologies Application to Multimodal Transportation. Developing and Implementing Advances to Transportation Practice,* http://www.ncgia.ucsb.edu/ncrst/synthesis/

EDM (2000) *Report on the Kocaeli Turkey Earthquake of August 17th 1999, EDM Technical Report No.* 6, EDM: Miki.

Eguchi, R., Houshmand, B., Huyck, C., Shinozuka, M and Tralli, D. (1999) A new application for remotely sensed data: Construction of building inventories using synthetic aperture radar technology, *MCEER research and Accomplishments1997-1999,* MCEER: Buffalo

Eguchi, R., Huyck, C., Houshmand, B., Mansouri, B., Shinozuka, M., Yamazaki, F. and Matsuoka, M. (2000a) The Marmara Earthquake: A View from Space, *The Marmara, Turkey Earthquake of August 17, 1999: Reconnaissance Report, Section 10, Technical Report MCEER-00-0001,* MCEER: Buffalo.

Eguchi, R., Huyck, C., Houshmand, B., Mansouri, B., Shinozuka, M., Yamazaki, F. and Matsuoka, M. and Ulgen, S. (2000b) The Marmara, Turkey earthquake: Using advanced technology to conduct earthquake reconnaissance, *MCEER research and Accomplishments1999-2000,* MCEER: Buffalo

Eguchi, R., Huyck, C., Adams, B., Mansouri, B., Houshmand, B., and Shinozuka, M. (2002) Earthquake damage detection algorithms using remote sensing data – Application to the August 17, 1999 Marmara, Turkey Earthquake, *Proceedings of the 7th U.S. National Conference on Earthquake Engineering,* Boston.

Eguchi, R., Huyck, C., Adams, B., Mansouri, B., Houshmand, B., and Shinozuka, M (2003) Resilient disaster response: Using remote Sensing technologies for post-earthquake damage detection, *MCEER research and Accomplishments 2001-2003,* MCEER: Buffalo.

Eguchi, R., Huyck, C. and Adams, B. (in preparation) *Advanced Technologies for Loss Estimation: Construction of Building Inventories Using Synthetic Aperture Radar and Optical Imagery,* MCEER Technical Report, MCEER: Buffalo.

ESA (2003) *Earth Watching,* http://earth.esa.int/ew/

Estrada, M., Kohiyama, M., Matsuoka, M. and Yamazaki, F. (2001a) Detection of damage due to the 2001 El Salvadore earthquake using Landsat images, *Proceedings of the 22nd Asian Conference on Remote Sensing,* Singapore.

Estrada, M., Matsuoka, M., Yamazaki, F. (2001b) Digital damage detection due to the 1999 Kocaeli, Turkey earthquake, *Bulletin of the Earthquake Resistant Structure Research Center,* 34, 55-66.

FEMA (2003a) *Mapping and Analysis Center: Remote Sensing Information and Data,* http://www.gismaps.fema.gov/rs.shtm

FEMA (2003b) *R&R Definitions*, http://www.fema.gov/rrr/conplan/appen_b.shtm

Fingas, M.F. and Brown, C.E. (1997) Review of oil spill remote sensing, *Spill Science and Technology Bulletin*, 4(4): 199-208.

Francis, P. and Rothery, D. (2000) Remote sensing of active volcanoes, *Annual review of Earth and Planetary Sciences*, 28: 81-106.

Galy, H.M and Sanders, R.A. (2000) Using SAR imagery for flood modeling, *Proceedings of the RGS-IBG Annual Conference*, Brighton.

Garofalo, D. and Wobber, F.J. (1974) A Nicaragua earthquake – Aerial photography for disaster assessment and damage, *Photographic Applications in Science, Technology and Medicine*, 9, 36-38.

Garshnek, V. and Burkle, F.M. (2000) *Communications and information tools for the 21st century: Changing the face of disaster response and humanitarian assistance*, http://www.ssgrr.it/en/ssgrr2000/papers/149.pdf

GEOMAC (2003) *GEOMAC Wildland Fire Support*, http://www.geomac.gov/

Hasegawa, H., Aoki, H., Yamazaki, F. and Sekimoto, I. (1999) Attempt for automated detection of damaged buildings using aerial HDTV images, *Proceedings of the 20th Asian Conference on Remote Sensing*, Hong Kong www.gisdevelopment.net/aars/acrs/1999/ts3/ts3097.shtml

Hasegawa, H., Yamazaki, F., Matsuoka, M. and Seikimoto, I. (2000) Determination of building damage due to earthquakes using aerial television images, *Proceedings of the 12th World Conference on Earthquake Engineering*, Aukland.

Hashitera, S., Kohiyama, M., Maki, N. and Fujita, H. (1999) Use of DMSP-OLS images for early identification of impacted areas due to the 1999 Marmara earthquake disaster, *Proceedings of the 20th Asian Conference on Remote Sensing*, Hong Kong, 1291-1296.

HAZUS (1997) *Earthquake Loss Estimation Methodology HAZUS: User's Manual*, FEMA.

Heuel, S. and Kolbe, T. (2001) Building reconstruction: the dilemma of generic versus specific models, *Künstliche Intelligenz*, 3, arenDTaP Verlag, Bremen.

Hiatt, M. (2002) Keeping our homelands safe, *Imaging Notes*, May/June: 20-23.

Huyck, C.K. and Adams, B.J. (2002) *Emergency response in the wake of the World Trade center attack: The remote sensing perspective*, MCEER Special Report Series, Volume 3, MCEER: Buffalo.

Huyck, C.K, Eguchi, R. and Houshmand, B. (2002) *Bare-earth algorithm for use with SAR and LIDAR digital elevation models*, MCEER-02-0004 Technical Report, MCEER: Buffalo.

Huyck, C.K, Mansouri, B., Eguchi, R.T., Houshmand, B., Castner, L. and Shinozuka, M. (2002) Earthquake damage detection algorithms using optical and ERS-SAR satellite data – Application to the August 17, 1999 Marmara, Turkey earthquake, *Proceedings of the 7th U.S. National Conference on Earthquake Engineering*, Boston.

Huyck, C.K., Adams, B.J. and Kehrlein, D.I. (2003) An evaluation of the role played by remote sensing technology following the World Trade Center attack, *Earthquake Engineering and Engineering Vibration*, 2(1): 1-10.

IDNDR (2003) *About the International Decade for Natural Disaster Reduction*, http://www.oneworld.org/idndr/frameset.html

IGOS (2001) *What is IGOS?*, http://www.igospartners.org/

International Charter (2003) *The Charter in Action*, http://www.disasterscharter.org/disasters_e.html

ISDR (2003) *International Strategy for Disaster Reduction*, http://www.unisdr.org/

Jayaraman, V., Chandreasekhar, M.G. and Rao, U.R. (1997) Managing the natural disasters from space technology inputs, *Acta Astronautica*, 40(2-8): 291-325.

JPL (1995) *The SRL Volcano Exhibit*, http://southport.jpl.nasa.gov/volcanopic.html

JPL (2003) *Glaciers and ice sheets in a changing climate*, http://www-radar.jpl.nasa.gov/glacier/

JRC (2003) *Natural Hazards Project*, http://natural-hazards.jrc.it/

Kerle, N. and Oppenheimer, C. (2002) Satellite remote sensing as a tool in lahar disaster management, *Disasters*, 26(2): 140-160.

Kohiyama, M., Hayashi, H., Maki, N. and Hashitera, S. (2001) *Night Time Damage Estimation*, http://www.gisdevelopment.net/magazine/gisdev/2001/mar/ntde.shtml

Laben, C. (2002) Integration of remote sensing data and geographic information system technology for emergency managers and their applications at the Pacific Disaster Center, *Optical Engineering*, 41: 2129-2136.

Langhelm, R. and Davis, B. (2002) Remote sensing coordination for improved emergency response, *Proceedings of the 15th Pecora Conference: Integrating Remote Sensing at the Global, Regional, and Local Scale*, Denver.

LaSen (2003) ALPIS: Airborne LIDAR Pipeline Inspection System, LaSen: Las Cruces.

Lavarake, A. (2001) How *GIS and remote sensing could have helped in the Gujurat disaster*, http://www.gisdevelopment.net/magazine/gisdev/2001/mar/hgrs.shtml

Logan, B. (2002) The lessons of 9/11, *Geospatial Solutions*, September: 26-30.

Lu, Z., Wicks, C., Dzurisin, D., Power, J., Thatcher, W. and Masterlark, T. (2003) Interferometric synthetic aperture radar studies of Alaska volcanoes, *EOM*, 12(2): 8-18.

Luscombe, B.W. and Hassan, H.M. (1993) Applying remote-sensing technologies to natural disaster risk management – Implications for developmental investments, *Acta Astronautica*, 29(10-11): 871-876.

Matsuoka, M. and Yamazaki, F. (1998) Identification of damaged areas due to the 1995 Hyogoken-Nanbu earthquake using satellite optical images, *Proceedings of the 19th Asian Conference on Remote Sensing*, Manila.

Matsuoka, M and Yamazaki, F. (2000a) Interferometric characterization of areas damaged by the 1995 Kobe earthquake using satellite SAR images, *Proceedings of the 12thWorld Conference on Earthquake Engineering*, Auckland.

Matsuoka, M. and Yamazaki, F. (2000b) Satellite remote sensing of damaged areas sue to the 1995 Kobe earthquake, In Toki, K. (ed) *Confronting Urban Earthquakes, Report of Fundamental Research on the Mitigation of Urban Disasters Caused by Near-field Earthquakes*, 259-262.

Matsuoka, M. and Yamazaki, F. (2002) Application of the damage detection method using SAR intensity images to recent earthquakes, *Proceedings of the IGARSS*, Toronto.

Matsuoka, M. and Yamazaki, F. (2003) Application of a methodology for detection building-damage area to recent earthquakes using SAR intensity images, *Proceedings of the 7th EERI US Japan Conference on Urban Earthquake Hazard Reduction*, Maui.

DHI (2002) *MIKE21 User Guide and Reference Manual*, Danish Hydraulic Institute: Hørsholm.

Mileti, D. (1999) *Disasters by Design: A Reassessment of Natural Hazards in the United States*, National Academies Press: Washington.

Mitomi, H., Yamazaki, F. and Matsuoka, M. (2000) Automated detection of building damage due to recent earthquakes using aerial television images, *Proceedings of the 21st Asian Conference on Remote Sensing*, Taipei.

Mitomi, H., Matsuoka, M. and Yamazaki, F. (2001a) Automated detection of buildings from aerial television images of the 2001 Gujurat, India earthquake, *Proceedings of the IEEE International Symposium on Geoscience and Remote Sensing*, Sydney.

Mitomi, H., Yamazaki F. and Matsuoka, M. (2001b) Development of automated extraction method for buildings damage area based on maximum likelihood classifier, *Proceedings of the 8th Conference on Structural Safety and Reliability*, Newport Beach.

Mitomi, H., Matsuoka, M and Yamazaki, F. (2002) Application of automated damage detection of buildings due to earthquakes by panchromatic television images, *Proceedings of the 7th U.S. National Conference on Earthquake Engineering*, Boston.

Morain, S. (2001) Remote sensing for transportation. Safety, hazards and disaster assessment, *Proceedings of the International Conference on Urban Geoinformatics*, Wuhan.

Nevatia, R and Price, K. (2002) Automatic and interactive modeling of buildings in urban environments from aerial images, *IEEE ICIP Rochester, NY*, Volume III: 525-528.

NDRD (2003) *Natural Disaster Reference Database*, http://ndrd.gsfc.nasa.gov/

NOAA (2003) *National Weather Service Radar Detects Columbia Space Shuttle Disaster*, http://www.crh.noaa.gov/ncrfc/News/news-index.html

Noronha, S. and Nevatia, R. (2001) Detection and modeling of buildings from multiple aerial images, *Transactions on Pattern Analysis and Machine Intelligence*, 23(5): 501-518.

Oberg, J. (2003) *High-tech sensor in the shuttle search*, http://www.msnbc.com/news/891627.asp?0sl=-13

Ogawa, N., Hasegawa, H., Yamazaki, F., Matsuoka, M and Aoki, H. (1999) Earthquake damage survey methods based on airborne HDTV, photography and SAR, *Proceedings of the 5th US Conference on Lifeline Earthquake Engineering, ASCE*, 322-331.

Ogawa, N. and Yamazaki, F. (2000) Photo-interpretation of buildings damage due to earthquakes using aerial photographs, *Proceedings of the 12thWorld Conference on Earthquake Engineering*, Auckland.

Profeti, G. and MacIntosh, H. (1997) Flood management through Landsat TM and ERS SAR data: A case study, *Hydrological Processes*, 11(10): 1397-1408.

Puzachenko, Y.G., Borunov, A.K., Koshkarev, A.V., Skulkin, V.S. and Sysuyev, V.V. (1990) Use of remote sensing imagery in analysis of consequences of the Armenian earthquake, *Mapping Sciences and Remote Sensing*, 27(2), 89-102.

Richards, P.B. (1982) Space technology contributions to emergency and disaster management, *Advances in Earth-oriented Applications of Space Technology*, 1(4): 215-221.

Roberts, D., Gardner, M., Regelbrugge, J., Pedreros, D. and Ustin, S. (1998) Mapping the distribution of wildfire fuels using AVIRIS in the Santa Monica Mountains, *Proc. 7th AVIRIS Earth Science Workshop JPL 97-21*, Pasadena.

Rodarmel, C. Scott, L., Simerlink, D. and Walker, J. (2002) Multisensor fusion over the World Trade Center disaster site, *Optical Engineering*, 41(9): 2120-2128.

RMSI (2003) Application of GIS for regional earthquake loss estimation, www.rmsi.com/PDF/regionalearthquake.pdf

Saito, K. and Spence, R.J. (in review) Using high-resolution satellite images for post-earthquake building damage assessment: a study following the 26.1.01 Gujurat earthquake, *Earthquake Spectra*.

San Miguel-Ayanz, J. Vogt, J., De Roo, A. and Schmuck, G. (2000) Natural hazards monitoring: Forest, fires, droughts and floods – The example of European pilot projects, *Surveys in Geophysics*, 21(2-3): 291-305.

SCWHC (2003) *The role of remote sensing*, http://www.crseo.ucsb.edu/resac/resac.html

Sharma, P.K., Chopra, R., Verma, V.K. and Thomas, A. (1996) Flood management using remote sensing technology: The Punjab (India) experience, *International Journal of Remote Sensing*, 17(17): 3511-3521.

Sinha, R. and Goyal, A. (2001) Lessons from Bhuj Earthquake, http://www.gisdevelopment.net/magazine/gisdev/2001/mar/lbe.shtml

Sun, W. and Sweeting, M (2001) An international disaster monitoring constellation with daily revisit employing advanced low-cost earth observation microsatellites, *Proceedings of the 22nd Asian Conference on Remote Sensing*, Singapore.

Tholey, N., Clandillon, S. and DeFraipont, P. (1997) The contribution of spaceborne SAR and optical data in monitoring flood events: Examples in northern and southern France, *Hydrological Processes*, 11(10): 1409-1413.

Thomas, D.S.K., Cutter, S.L., Hodgson, M., Gutekunst, M. and Jones, S. (2002) *Use of spatial data and geographic technologies in response to the September 11 terrorist attack*, http://www.colorado.edu/hazards/qr/qr153/qr153.html

Tralli, D.M. (2000) *Assessment of Advanced Technologies for Loss Estimation*, MCEER: Buffalo.

Walter, L.S. (1990) The uses of satellite technology in disaster management, *Disasters*, 14(1): 20-35.

Walter, L.S. (1991) The role of space technology in disaster mitigation, *Workshop on the Application of Space Techniques to Combat Natural Disasters*, Beijing.

Wang, Q., Watanabe, M., Hayashi, S. and Murakami, S. (2003) Using NOAA AVHRR data to assess flood damage in China, *Environmental Monitoring and Assessment*, 82(2): 119-148.

Werner, S., Lavoie, J., Eitzel., C., Cho, S., Huyck, C., Gosh, S., Eguchi, R., Taylor, C. and Moore, J.E. (2003) REDARS 1: Demonstration software for seismic risk analysis of highway systems, *MCEER research and Accomplishments 2001-2003*, MCEER: Buffalo.

Western Governors Association (2000) *Utility and use of GIS and remote sensing technologies to support disaster services role at the local state and federal level, Policy resolution 00-034,* http://www.westgov.org/wga/policy/00/00034.pdf

Williamson, R. A. and Baker, J.C. (2002) Lending a helping hand: Using remote sensing to support the response and recovery operations at the World Trade Center, *PE&RS*, 68(9): 870-896

Williamson, R., Morain, S., Budge, A. and Hepner, G. (2002) *Remote Sensing for Transportation Security, Report of the Washington NCRST Workshop,* NCRST-H: Albuquerque.

Xiong, D. (2003) *Automated Road Network Extraction from High Resolution Images*, NCRST-H White Paper, http://riker.unm.edu/DASH_new/pdf/White%20Papers/Automated%20Road%20Network%20Extraction.pdf

Yamazaki, F. (2001) Applications of remote sensing and GIS for damage assessment, *Proceedings of the Joint Workshop on Urban Safety Engineering*, Asian Institute of Technology, Bangkok.

Yusuf, Y., Matsuoka, M. and Yamazaki, F. (2001a) Damage assessment after 2001 Gujrat earthquake using Landsat-7 satellite images, *Journal of the Indian Society of Remote Sensing*, 29(1), 233-239.

Yusuf, Y., Matsuoka, M. and Yamazaki, F. (2001b) Damage detection from Landsat-7 satellite images for the 2001 Gujurat, India earthquake, *Proceedings of the 22nd Asian Conference on Remote Sensing*, Singapore.

Yusuf, Y., Matsuoka, M. and Yamazaki, F. (2002) Detection of building damage due to the 2001 Gujurat, India earthquake, using satellite remote sensing, *Proceedings of the 7th U.S. National Conference on Earthquake Engineering*, Boston.

CONTEXT AND RESILIENCY:
INFLUENCES ON ELECTRIC UTILITY LIFELINE PERFORMANCE

DOROTHY REED[1], JANE PREUSS[2] and JAEWOOK PARK[3]

ABSTRACT

The project reported in this paper documents the performance of an urban electric power system located in the Pacific Northwest region of the US for five events: four winter storms and the Nisqually earthquake of magnitude 6.8. In order to assess the performance of this power system for these events, several metrics were employed. These included linear regression of weather variables with standard utility reliability indices as defined by IEEE (2001); restoration rate following the initial impact of the event; distribution of the duration of outages; and the empirical derivation of a fragility curve for the lifeline comprised of feeders, lines, poles and substations as a function of seismic parameters. A qualitative investigation of local policies provides limited evidence of local policy correlation with performance. The preliminary findings suggest that certain metrics are more useful than others when making comparisons of performance for more than one hazard.

INTRODUCTION AND BACKGROUND

Electric utility lifelines are comprised of three subsystems: generation, transmission and distribution. The elements of the system are illustrated simplistically in the diagram of Figure 1. [*Distribution System Reliability*, (2002)].

Typically, reliability efforts are focused upon one of the three subsystems. For example, distribution systems are subject to particular scrutiny and standards for reliability indices have been established by IEEE. These are SAIDI (System Average Interruption Duration Index) and SAIFI (System Average Interruption Frequency Index). They are defined as follows [IEEE, (2001)]:

$$SAIDI = \frac{\sum \text{Customer hours off for each interruption}}{\text{Total number of customers served}} \quad (1a)$$

$$SAIFI = \frac{\sum \text{Customers affected by each interruption}}{\text{Total number of customers served}} \quad (1b)$$

Limitations of these measures of reliability include the following: The definition of "outage" varies from utility to utility. Some do not consider outages lasting less than a minute, while others do not consider outages lasting less than five minutes in their calculation. In addition, a business may be counted as one customer whereas a household is also counted as a customer. Also, because outages may "travel" from one region to another during an event, the exact number of people affected is not truly reflected in the indices since they are not evaluated spatially or over time.

---

[1] Prof.of Civil and Environmental Engineering, University of Washington email: reed@u.washington.edu
[2] Principal GeoEngineers, Inc email: jpreuss@geoengineers.com
[3] Graduate Research Assistant, Civil and Environmental Engineering, University of Washington.

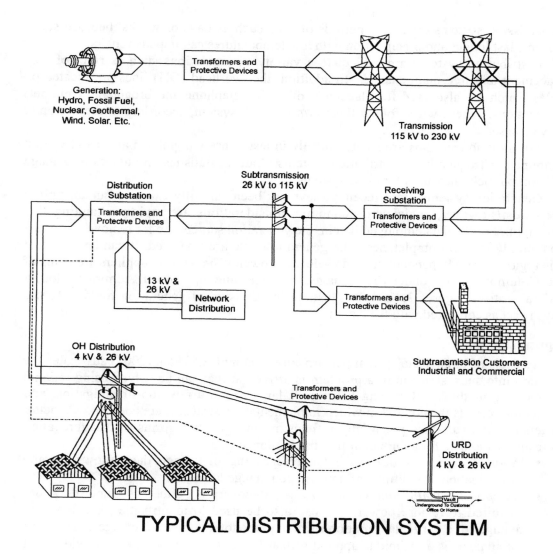

TYPICAL DISTRIBUTION SYSTEM

Figure 1. Simplified diagram of a typical US utility lifeline. Taken from [Seattle City Light, *Distribution System Reliability*, (2002)].

Voltage drops from transmitted electricity to distribution systems generally occur at various substations. These substations have been the focus of attention since they tend to fail in shallow-focus earthquakes typical of California events. In contrast, during the comparatively deep focus Nisqually event of 2001, the Pacific Northwest region did not experience prolonged outages due to substation failures. Unfortunately, the causes of each failure were not recorded during the repair process so that the exact nature of the damage is not known. Weather-related damage occurs to both the distribution and transmission systems over long periods of time such as days or weeks, because storm systems last longer and precipitation with low temperatures can impede repairs.

Distribution systems are subject to a variety of regulations such as right of way placement and tree maintenance. In addition, the overhead (OH) lines are located on poles which are also used for placement of cable, telephone and broadband (internet) connections. These wires follow their own network system, according to the customer service base.

Transmission systems are located mainly in less densely populated areas and are not examined as frequently as distribution systems. Indeed, statistics on outages or damage are not routinely measured for these systems.

Generation systems, until recently, have not been carefully examined under failure scenarios; however, social and economic considerations are rapidly changing this oversight. For the data employed herein, the generation sources are far removed from the community but for completeness, the generation data are reviewed. It can be seen that this region is very dependent on hydro-electric power. The largest suppliers are dams of the Columbia River, under the management of Bonneville Power Administration, a federal entity. Figures 2a and 2b illustrate the division of generation for the Pacific Northwest as compared with the nation.

APPROACH

Analysis of large-scale system performance is complex. Petak (2003) advocates the division into subsystems until a problem is arrived at that can be solved using rational engineering methods. Reducing risk is to a large extent a function of mitigation, and mitigation reflects the context in which implementation measures are adopted. To better understand the variables potentially influencing outages, two forms of disaggregation were applied to data for an urban utility distribution system:

- *Spatial extent of damage.* Georeferencing of outage data allows for the investigation of multiple influences upon outage location.
- *Temporal extent of outage or duration.* Recovery is defined as "time (duration) required to restore power." It can also be used to identify the percentage of outages of a particular duration relative to the total number of outages. This type of analysis is a technique applied primarily for wind analysis, e.g., Davidson, et al. (2003) and Reed, et al. (2002a,b); however, outage duration analysis is a tool that applies equally well to both wind and seismic events. In contrast, restoration is the percentage of the total network that has had power restored. One hundred percent restoration means that all power has been restored. Restoration rates following earthquakes have been under investigation for many years, e.g., Nojima, and Sugito (2002) collected restoration data following the 1995 Great Hanshin-Awaji earthquake.

Table 1 describes the five disaster events investigated in this paper. To investigate differences in spatial and temporal differences in outages a dual focus comparative analysis was applied to the utility distribution system, which was disaggregated on the basis of sub-areas that experienced differences in outage durations. Context, in the form of local differences in policies, was then applied to the disaggregated areas. Traditional engineering approaches for fragility and vulnerability were also undertaken.

Table 1: Case Study Events

Date	Type of Event	Event duration	Number of feeders affected	Feeders affected as % of total feeders in system
January 1993	Winter storm	63.1 hours	107	45.9
November 1995	Winter storm	12.7 hours	12	5.2
November 1996	Winter storm	30 hours	39	16.7
December 1996	Winter storm	143.7 hours	34	14.6
February 2001	Earthquake (Magnitude 6.8)	30-40 seconds	17	10.7

Outage data were obtained from field logs that identified outages by location (feeder) and cause. ***The security agreement for the use of these data prevents publishing the feeder file.*** The distribution system of the Seattle area that was studied is bounded on the western side by Puget Sound and by Lake Washington on the east. Portions of the system lie outside the Seattle city boundaries. The exact locations of the underground lines are not provided; the majority of the system is above ground. The system is comprised of 230 feeders for a total length of over 3543 km. Six substations are contained in this system. The length per feeder is not constant; the maximum feeder length covers approximately 58 km, while the minimum is limited to approximately 12 m. The *Arcview (GIS)* file of the system consists of over 25,000 poly-lines with the maximum number of line segments per feeder being 473.

To establish a basis for analyzing the implications of locally based policies and regulations, data were georeferenced to convert the feeders to street locations. In this way data pertaining to feeders could be correlated with communities located within the utility's boundaries and with geologic characteristics such as liquefaction areas. It should be noted that the correlation was not perfect because feeders do cross both of these boundaries; i.e., jurisdictional city limits and the outer edges of the liquefaction areas.

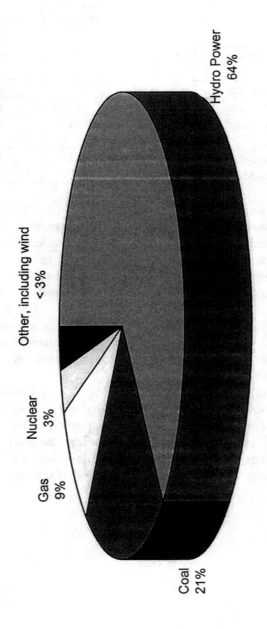

Figure 2a. Power generation sources for the region that includes Washington, Oregon, Idaho, Montana and parts of Wyoming (Data derived from the Seattle Times)

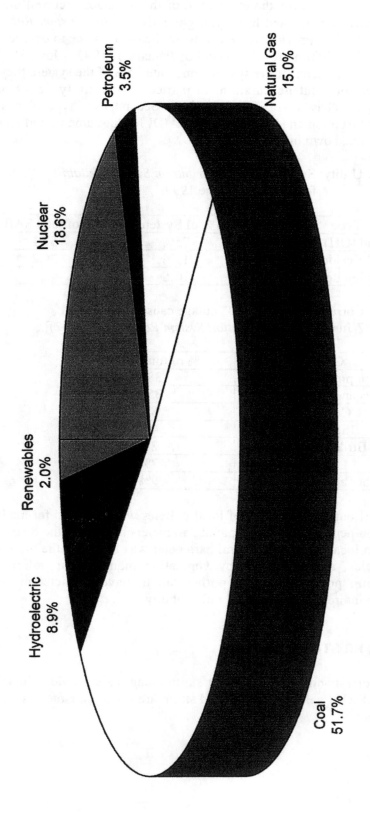

Figure 2b. US Electric Industry Net Generation (Source: Energy Information Administration; EIA(2001))

In order to provide a context for the evaluation of the damages incurred during a typical period, data provided by Seattle City Light [*Distribution System Reliability* (2002)] are used. First, Table 2 provides an overview of the outages due to overhead and underground portions of the lifeline for the period of January 1, 1997 – June 30, 1999. Although the overhead components of the system constitute 75% of the system they only account for 51% of the total outage duration. In particular, the utility tracks outage according to eight categories based on IEEE regulations. Second, as a "typical" baseline, the causes of system interruption and the associated SAIDI for the same period (January 1, 1997-June 30, 1999) are shown in Table 3.

Table 2: Case Study Utility: System Type [*Distribution System Reliability*, (2002)] for Jan 1997 – June 1999.

System Type	% of System	% of total SAIDI
Overhead (OH)	75%	51.1
Underground Residential	15%	36.2
Network (selected high density areas)	10%	.5

Table 3: Correlation of "typical" outage causes with SAIDI for Jan. 1997-June 1999 [*Distribution System Reliability*, (2002)].

Cause	% of total SAIDI
Equipment Failure	38%
Trees and Wires down	35%
Car/pole	8%
Lightning	4%
Birds/animals	2%
Dug up	2%
Other and unknown	9%

Based on patterns of outage, two types of local policies were selected for analyzing the effects of context on performance. The spatial parameters addressed the distribution of outages according to location. The temporal parameter was based on the duration of outages. Local policies, as represented by vegetation management policies and comprehensive planning policies, were hypothesized to have impacts on system interruption indices and indirectly on structural vulnerability.

CASE STUDY: WINTER STORM
Weather Data
The weather data corresponding to the winter storm outages are provided in Table 4 below. The SAIDI and SAIFI values for a selected storm are usually denoted as STAIDI and STAIFI, respectively.

Table 4: Storm Data Summary

OUTAGE DATA	1/93 Storm	11/95 Storm	11/96 Storm	12/96 Storm
Total storm duration in hours	63.08	12.72	30	143.65
Total number of customers in the distribution system	344000	365475	348296	348296
Number of feeders affected	107	12	39	34
STAIFI [interruptions/year]	75.336	18.396	29.784	7.884
Number feeders affected/total number of feeders in system	0.459	0.052	0.167	0.146
STAIDI [hours/year]	823.5	37.9	110.4	40.1
Storm Weather Data				
Minimum Temperature [°C]	-1.12	7.84	-2.24	-2.8
Maximum 24 hour precipitation [cm]	2.032	0.4318	1.1176	6.985
Maximum 2-minute windspeed during the storm measured for standard US conditions [m/s]; [ASCE7-02]	20.54	16.55	9.7	9.7
Corresponding peak 5-second gust [m/s]	31.31	19.01	13.12	15.98
Base Values for the Reliability Indices				
SAIDI [hours/year]	26.96	29.31	36.3	36.3
SAIFI [interruptions/year]	0.28	0.32	0.57	0.57

The data show that the most severe storm from an outage point of view was the January 1993 storm in which almost half (45.9%) of the system was affected. The December 1996 storm was marked by large accumulations of ice and snow at low temperatures. Available wind data from three Seattle locations consistently show a strong SW-NE wind flow during the storm for the December 1996-January 1997 event. The areas of primary outages for this extended storm were clustered at the north and south ends of Lake Washington and not along heavily exposed Elliott Bay. Clearly, local topography plays an important role in performance of the feeders.

Reliability Index Analysis

Using the data for all four storms, Reed and Cook (1999) used linear regression to identify three models. Because they were identifying SAIDI for particular storm events, rather than a composite for the year, they used the notation STAIDI and STAIFI to indicate that the storm data were employed. They found that when normalized by base values for non-storm SAIDI and SAIFI during the years the storms occurred, respectively, the following best-fit models are

$$STAIFI_n = -38.5 + 6.67G^2 \text{ for an } R^2 = 94.1\%$$

$$STAIDI_n = -8.50 + 0.841G^2 + 0.071P \text{ for an } R^2 = 91.9\%$$

and

$$STAIDI_n = -10.8 + 0.841G^2 + 2.08T \text{ for an } R^2 = 96.1\%$$

where

$STAIDI_n$ is the ratio of STAIFI[int/yr] to SAIFI[int/yr]; $STAIDI_n$ is the ratio of STAIDI[hr/yr] to SAIDI[int/yr]; G is the ratio of the peak storm gust to the monthly peak; P is the ratio of the storm precipitation to the monthly maximum; and T is the ratio of the difference between the monthly minimum and the storm minimum to the monthly minimum. The use of the squared normalized sustained windspeeds for predicting $STAIDI_n$ only produced an R^2 of 62.4%. Clearly the peak gust squared is the best predictor. The SAIFI result is consistent with a model predicted by Brown, et al. (1997a,b,c). The SAIDI correlations may reflect the difficulty of repair crew travel to outage locations under heavy ice conditions as well as structural failures of the subsystem. These relationships are limited to distribution system failures and do not take into account any relationships with transmission system failures that may have occurred. In order to evaluate what types of engineering solutions to the storm-related damage might exist, the manner in which the weather wreaked havoc was investigated.

Recorded Causes of Outages

  The causes of the outages as described by the repair crews are shown in the pie charts of Figure 3.

*The category "components" combines fuses, sectionalizers, jumpers, transformers, etc. and "unknown" combines "storms" and "wires down".

Figure 3. Causes of Storm Outages for the November 1995, November 1996 and December 1996 Events.

  In most instances, the true cause or nature of a fallen line or outage is not known or recorded. "Unknown" includes animal-related failures such as those created by birds or squirrels. Most crews concentrate on fixing the power system, rather than noting causes of outages in detail. The inexact nature of this data collection presents difficulties for structural engineers because the degree to which the various system components can be redesigned or modified to result in prescribed structural damage more easily repaired than at present is not obvious.

Outage Duration Analysis

For winter storms, the weather conditions vary over a period of hours and even days. The conditions may adversely affect the restoration efforts as well as contribute to additional outages. Rather than using restoration curves, the analysis of duration is used for wind conditions. Following the approach of Davidson, et al. (2003) the storms were analyzed as shown in Figure 4, where the number of outages less than or equal to a duration in hours is plotted for each storm. Preliminary analysis of the probability distribution for these curves indicates a lognormal fit. The Nisqually data are shown for comparison and they will be discussed in the earthquake section in greater detail. The challenges of using probability distributions for these data are acknowledged; i.e., one outage may be affected in its duration by other outages. For example, if a finite set of crews are available, outages are "repaired" in an order based upon several factors such as geographical location, time of day, weather conditions, electrical network configuration, and other unknown variables.

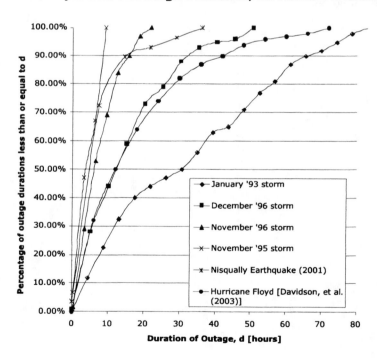

Figure 4. Outage comparison for all events excluding flickers. (A flicker is an outage of less than one minute). The Hurricane Floyd data are taken from Davidson, et al. (2003).

Local Area Evaluation: Policy Analysis

Initial spatial analysis showed that certain areas experienced outages of longer duration than others or more frequently than others. Because the weather data were not available for each subsection, but rather only recorded at select meteorological stations, a localized study was not possible. However, the data recorded by the utility in terms of "causes" was investigated locally. One aspect of this local analysis was to evaluate any differences in policies that might affect the outage durations or frequencies.

The first step toward identifying the possible influence of local policies on outages was to correlate system disruption with noted cause. Policy issues were investigated for those events that had adequate data. For the 1995 and 1996 wind storms, the data indicate that of the eight IEEE categories, the highest percentage was from trees for overhead systems (75% of the system), and equipment failure for underground elements.

The utility has organized statistics on the distribution system performance by sub-areas. Review of the storm outage date from the winter storms consistently found that sub-areas within the urban area experienced consistently lower rates for the categories "trees/wires down" than the communities outside the major jurisdiction. For example the southwest sub-area experienced 24% SAIDI disruption from "tree/wires" while the southeast experienced 16%. The sub-areas north of the city experienced 68% of SAIDI disruption from "trees/wires" while the south experienced 45%. This discrepancy prompted the investigators to hypothesize that differences in SAIDI could be influenced by differences in local policies implemented by the various jurisdictions. Figure 5 indicates differences in outage times during the 1996 winter storm and the distribution of those times by subareas.

Figure 5: December 1996 Wind Storm: Outage Durations

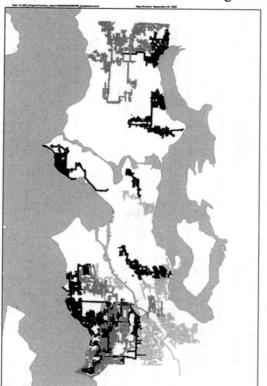

Explanation

December 1996 Wind Storm Outages

    <3 hrs

    3 hrs - 12 hrs

——— >12 hrs w/max of 36 hrs

Further disaggregating of the utility service area revealed that there are eight incorporated municipalities within the limits of the area serviced by the utility. Each jurisdiction adopts and enforces its local codes and policies; although, two of the smaller jurisdictions use the County's codes. Therefore, for this analysis there are a total of seven different local policies which establish the planning and regulatory context.

Tree related causes constitute the cause of the majority of reported outages. Vegetation management policies were therefore selected as the form of local policies to evaluate for their possible influence on reliability. All of the jurisdictions have vegetation management policies; however, the underlying intent of the policies and specific provisions differ. For the seven jurisdictions five key variables were analyzed as illustrated in Table 5.

Table 5. Local Policy Analysis for Winter Storm Data

Variable	Key Elements	Number of jurisdictions affected	Comments
Legal Status	Adoption by code Adoption as policy	1 6	Some jurisdictions adopted vegetation management guidelines, which are advisory (not legally binding) in terms of status. Only one jurisdiction adopted portions of the vegetation management policy by a code (legally enforceable). This jurisdiction with more stringent enforcement capability has a significantly lower rate of tree-related outages.
Date adopted	1995 and earlier 2000 2001-2002	1 2 4	It takes time for vegetation management policies to positively achieve their desired results. The length of time that the jurisdiction had a policy to enforce was therefore noted. The jurisdiction that had adopted the code prior to 1995 is the same jurisdiction that had time to gradually replace hazardous trees with species that are less prone to either breaking or falling onto the wires.
Prohibited trees	Prohibits specific species No limitations	1 6	The code for the single jurisdiction prohibits the planting of three species which are brittle, large, or have invasive root systems that can interfere with the utility's system. This jurisdiction also has a list of species that are recommended for planting within a specified distance of power lines. Analysis of outages during the major storm events revealed that the lowest outage duration in all of the events was found in the jurisdiction that had adopted the list of prohibited species.
Tree Removal	Removal provisions No removal provisions Removal prohibited	4 1 2	Four jurisdictions require removal of "danger trees" which are diseased or potentially hazardous. One jurisdiction has no tree removal provisions. It is particularly interesting to note that two jurisdictions prohibit removal of trees; thus, vegetation management to remove trees before they fall onto power lines is made extremely difficult—if not impossible. Such ordinances effectively increase the hazard.
Setbacks	From utility lines None required	2 5	The setbacks specify how far trees must be planted and trimmed back when located in public areas. The two communities that identify setbacks adopt the policy as guidelines rather than by code.

CASE STUDY #2: NISQUALLY EARTHQUAKE

On February 28, 2001, a magnitude 6.8 earthquake struck western Washington State in the US. The epicenter was approximately 58 km southwest of Seattle. Initial estimates by the United States Geological Survey indicated a focal depth of 60 km. The utility provided system wide outage statistics for the Nisqually earthquake. However, unlike the wind storms, specific causes of failures were not documented by the repair crews who simply noted "earthquake" rather than pinpointing components of the system that were damaged. Outages during the Nisqually earthquake event were initially correlated with several parameters including areas of uncompacted fill and regions of large Modified Mercalli Intensity values as shown in Figure 6. Variables influencing outages are as described in the following sections.

Figure 6: Nisqually Earthquake Intensity

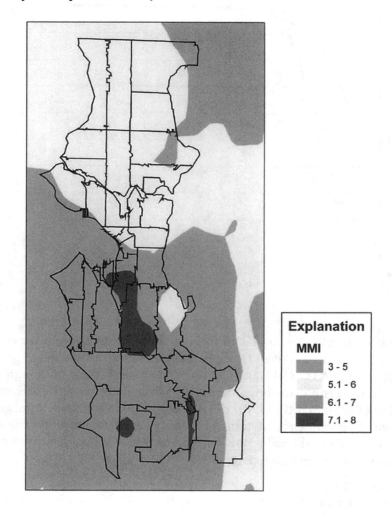

Restoration Rate

Restoration efforts were swift and concluded within thirty-six hours after the event. Figure 7 shows the actual restoration and the best-fit rate R as defined by [e.g., Chang, (1998)]:

$$R = 1 - e^{-bct} \tag{2}$$

where b = 10.06; c =1 for time t in days. In comparison, for lifelines in the Bay area following the Loma Prieta earthquake, b = 2.75 and c =1. The relatively high value for the parameter b in this study reflects the rapid recovery for the Seattle area.

Figure 7. Restoration rate for the Nisqually Earthquake.

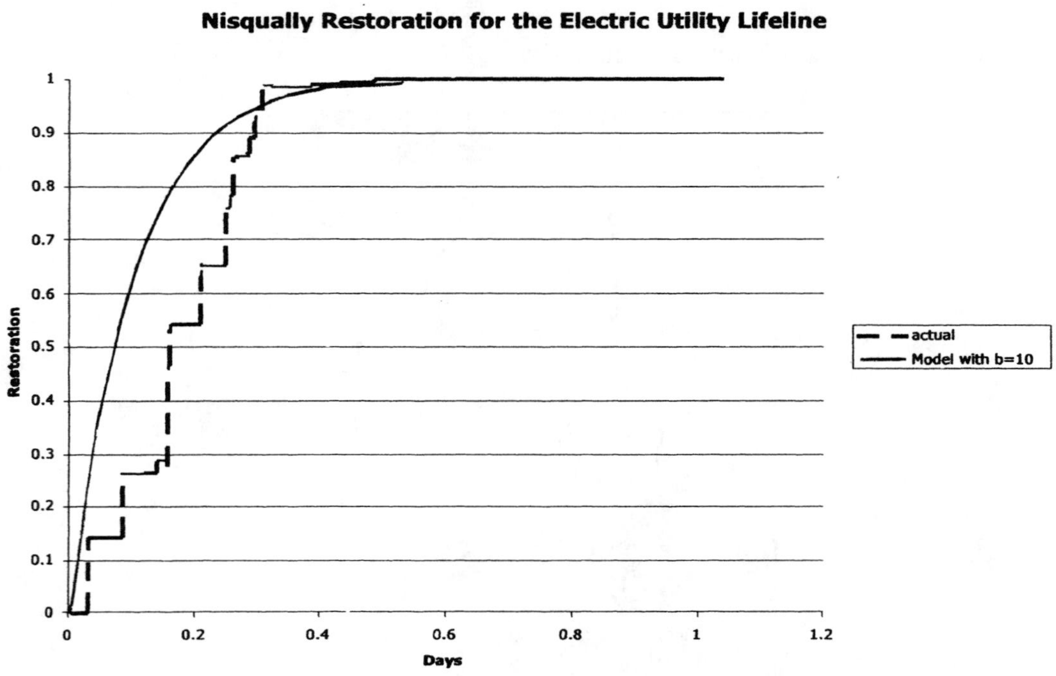

Nisqually Restoration for the Electric Utility Lifeline

- - actual
— Model with b=10

Restoration

Days

The restoration results are similar in form to those obtained by Nojima, and Sugito (2002) for the Great Hanshin-Awaji (Kobe) earthquake of 1995 and in the general format provided by ATC-25 [ATC (1991)]. It is noted that substations, although historically more vulnerable to seismic loadings during California events [e.g., Anagnos (1999); Hwang and Huo (1995); Schiff (1998)] did not experience significant damage during the Nisqually event. The influence of transmission system damage upon the distribution system failure was not considered in this analysis as transmission data were not available for consideration. There were no reports of generation failure.

Outage Duration Analysis

In addition to an analysis of restoration, the probability distribution of outage durations as shown in Figure 8 was investigated. Of the recorded outages, twenty-four percent were recorded as "flickers", that is, outages of duration less than or equal to one minute or sixty seconds. In this analysis of the outage durations, the "flicker" data were eliminated from consideration.

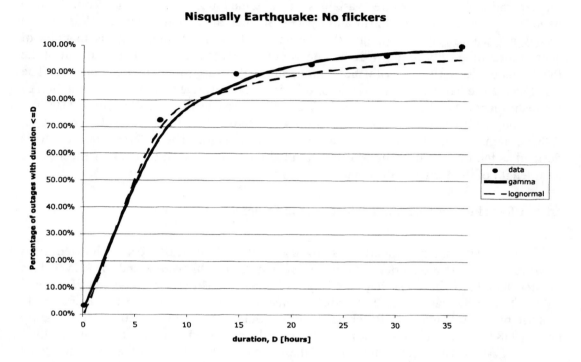

Figure 8. Outage duration analysis for Nisqually.

Nojima, and Sugito (2002) found that a Gamma distribution best fit outage data from the Kobe event which included outages for electricity, water and gas lifelines. A lognormal distribution was a better fit for the Nisqually electricity data; however, it is noted that the Nisqually data set, with the maximum outage duration of approximately thirty-six hours, is much smaller than for the more severe Kobe event.

<u>Fragility Curves</u>

Evaluation of the Damage Index

All of the outage duration data were examined for correlations with peak ground velocity (PGV), peak ground acceleration (PGA) and instrumental intensity (MMI). GIS representation of the latter three variables was obtained in *ArcView* format from the web page of the Department of Earth and Spaces Science at the University of Washington (www.ess.washington.edu/shake/). The value of the PGA shape file is in units of acceleration of gravity or g's with the interval of the contour being 0.04g. The unit of the PGV shape file is cm/sec with the contour interval of approximately 2.0 cm/sec. The MMI shape file has a contour interval of 0.2 intensity units. Details of the computation of the instrumental MMI as based upon PGV and PGA are described in Wald, et al. (1999).

Through GIS-based comparisons similar to those described by O'Rourke, et al. (2001), damage was evaluated through an intensity factor. The *Damage Index* (DI) is defined as the ratio of the length of the affected feeders in a specific contour level to the total length of the feeder in the contour level, i.e.,

$$DI_i = \text{affected feeder length in contour i / total feeder length in contour i} \qquad (3)$$

The Damage Index comparison showed that about half of the feeders located in an MMI region of 7 to 8 experienced outages and similar results were found for PGV values greater than 22 cm/sec. The region of highest damage was located in the waterfront and adjacent upland areas of Seattle that have a history of unconsolidated fill placement. In this vicinity, the Duwamish River has had the course of its flow modified significantly since 1908 [Tanner, (1991)]. Prior to that date fill was placed along the margins of Elliott Bay. The MMI was greatest for both of these historic areas built on the unconsolidated fill. Because the utility network in this region is comprised mostly of lines and poles rather than substations, it is assumed that poles are more vulnerable to base movement as described by ground velocity than acceleration. This result will be investigated further if more data become available for other earthquake events.

Logistic Regression Modeling

The damage index values were used to develop fragility curves. Based upon previous studies by Nojima and Sugito (2002), the logistic regression model was used to model the fragility. That is, the model of the conditional expectation $E(Y|x)$, where Y is defined as the damage index DI and x is the earthquake parameter such as the instrumental MMI, PGV or PGA, is sought.

The logit transformation is defined as $\pi(x) = E(Y \mid x)$ to represent the conditional mean of Y given x when the logistic distribution is used [e.g., Hosmer and Lemeshow, (2000)]. The form of the logistic regression model used is

$$\pi(x) = \frac{e^{\beta_0 + \beta_1 x}}{1 + e^{\beta_0 + \beta_1 x}} \qquad (4)$$

The logit transformation results in

$$g(x) = \ln\left[\frac{\pi(x)}{1 - \pi(x)}\right] = \beta_0 + \beta_1 x \qquad (5)$$

Using the statistical software package *S-Plus*, this model was fit to the Nisqually data to yield the following tabulated results:

Table 6: Power System Models

Nisqually parameter	β_0	β_1
$\text{Log}_{10}(\text{PGA [in g's]})$	3.4544	6.9240
$\text{Log}_{10}(\text{PGV [cm/sec]})$	-10.8614	7.6203
Instrumental MMI	-17.1017	2.3824

The fragility curves are plotted in Figures 9a-c.

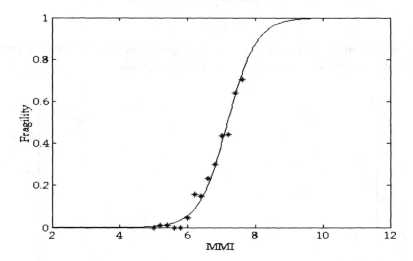

Figure 9a. The logit model (-) for the utility damage index or fragility as a function of instrumental MMI compared with the raw data (+) for Nisqually.

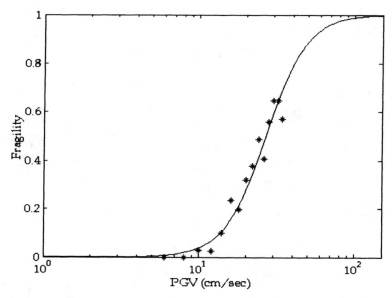

Figure 9b. The logit model (-) for the utility damage index as a function of peak ground velocity (PGV) compared with the raw data (+) for Nisqually.

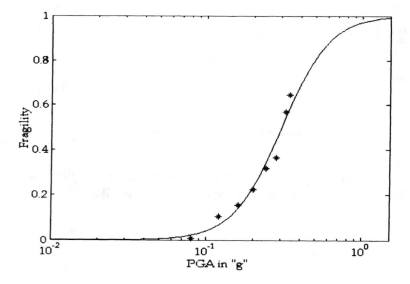

Figure 9c. The logit model (-) for the utility damage index as a function of peak ground acceleration (PGA) compared with the raw data (+) for Nisqually.

Policy Analysis

Because outage causes were not recorded for the Nisqually event, the policy analysis process was "reversed" from the winter storm investigation. That is, consequences of outage were evaluated by examining the characteristics of uses that were disrupted.

Those uses were then analyzed in relation to underlying comprehensive planning and zoning policies. Business interruption data provided by Meszaros and Fiegener (2002) were used extensively. According to Meszaros and Fiegener (2002):

> "Nisqually was notable for how well the region's lifelines performed. Gas and water were mostly undisrupted. Electricity failures were somewhat more common. Disruptions to communication lines were the most common and most long lasting."

Their conclusions are based on a survey mailed to 4,000 randomly selected Puget Sound members of the National Federation of Independent Businesses (NFIB) approximately nine months after the earthquake. The survey 862 respondents were asked whether they experienced any electricity disruptions to normal business operations as a result of the earthquake. If they replied "yes" respondents were asked the duration of the outage. The responses to the electricity question are summarized in Table 7:

Table 7: Nisqually Earthquake: Business Interruption Because of Electricity

No Disruption	<2 hours	Most of 1st day	Part of 2nd day	3-6 days	>1 month	Total
732	39	76	11	3	1	862

The business interruption data were georeferenced by zip codes. The zip code based community internet MMI values were therefore used for correlation studies rather than the instrumental MMI values (shown in Figure 6) employed for the fragility study. (The instrument based MMI varied slightly from the MMI values achieved through an internet survey). Not surprisingly the largest concentration of interruptions was found to center around the area with the highest recorded community internet MMI and the area with the greatest liquefaction as shown in Figure 10.

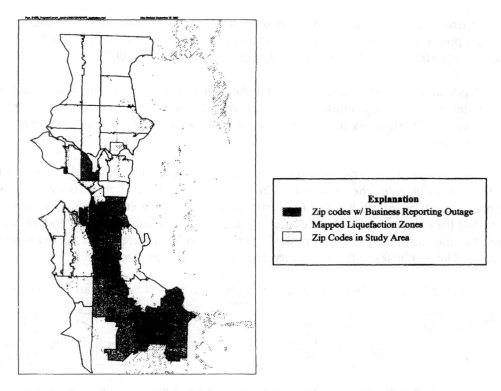

Figure 10: Business Interruptions from Electricity correlated with Liquefaction

Characteristics of the locations where no business interruptions were reported were then compared with the locations that reported interruptions. The areas with the highest density of business and commercial use are in the central business district and the university district. Both are areas in the city with the highest redundancy ("network system" indicated in Table 2) that only incurs .5% of the utility's SAIDI. There were no reported outages in these two areas (either because there were no respondents or no outages). Conversely the areas south of downtown have lower density development; in addition buildings tend to be older—including historic areas and industrial areas. Thus business interruption correlates with a number of interrelated variables such as older buildings and infrastructure, as well as old uncompacted fill that liquefied during the earthquake. The survey data unfortunately did not give insights into whether electricity outages resulted from failures in the distribution systems, or from damage to the building. There were however a number of responses reporting "no electricity" in areas where it is known that there was no problems on that feeder; in such cases, it is assumed that the problems stemmed from building damage.

The next step is to learn more about the underlying comprehensive plan and zoning policies. Although this analysis is currently underway, it is hypothesized that the zoning for these areas south of the central business district are designated "medium density commercial" and "industrial". Structures in such zones tend to be smaller and lower-rent than in the higher density areas. Small businesses tend to cluster in these lower density areas. It should be noted that 86.9% of the population in that area had fewer than 20 employees, while 81.6% of respondents had less than 20 employees [Meszaros and Fiegener (2002)].

CONCLUSIONS
Rates of Restoration by Hazard

Restoration rates differ by hazard. For wind, damage is sequential over the duration of the storm since trees or branches may fall at different points of the feeder multiple times, with the requisite need for repair. For earthquakes, most damage occurs during the first event and despite aftershock activity is usually not subject to repeated interruption. Outage durations are best compared through the format as shown in Figure 4. One of the leading software tools for earthquake damage prediction, HAZUS-99 [FEMA (2000)], does not consider distribution systems, but rather it bases all outage information on damage to transmission and generation facilities. This scenario may be appropriate for shallow-focus events, such as those historically occurring in California, but not for the deep-focus Nisqually event that occurred in Washington State. The results presented here suggest that further investigations of urban outages predictions may yield insight into distribution system behavior independent of transmission system performance.

Context is Important but Varies by Hazard

For wind, the socio-political context is best represented by vegetation management policies. These policies were selected based on review of outage causes given to the project by the participating utility. These policies appear to influence both the spatial and temporal distribution of outages. Spatially, those jurisdictions that had adopted regulations prohibiting specified "danger" species experienced dramatically lower outage occurrences and durations than those jurisdictions that did not have ordinances with regulatory authority. It should also be noted that selected jurisdictions had vegetation management ordinances which prohibited the cutting of trees. Utilities have more difficulties removing and trimming "danger trees" in these communities because a permit is required, which can (and often is) denied.

For the earthquake scenario it was more difficult to establish the socio-political context because specific components of the system resulting in the outages were not documented. Instead the project qualitatively correlated consequences (business interruption) with system characteristics and subsequently with comprehensive plan and zoning designations. As indicated in Table 2 the most robust system type (accounting for 10 percent of the utility service area) is in the central downtown core and university. Although these areas did not experience as high a ground motion as the area to the south it should be noted that no respondents returned questionnaires—or reported outages in these areas. It is not unreasonable to assume that there were negligible outages in these areas. Conversely, from a policy context, it is also reasonable to assume that land use drove the utility to ensure higher reliability to the two high density areas that constitute the center of the region's commercial, education, and research activities.

In summary, based upon limited data evaluation, local policies appear to influence reliability; however, the specific policies are different under winter storm vs. earthquake conditions.

Risk Reduction is Multidisciplinary

Reduction of risk from earthquake and wind events is a problem that typically tends to be approached from a technical design focus. The influences of such non-technical policies of vegetation management and land use suggest the need to develop integrative approaches. Such approaches must combine the insights of interdisciplinary perspectives, while finding gaps in knowledge and understanding between disciplines. The first step in this integration is to further disaggregate the disruption in terms of time and space. Then the ranges of influencing policies can be further refined and alternative perspectives on contextual modifications considered. The integrative approach is also an important way to educate decision makers concerning the contribution that their current ordinances may make to outages.

In many instances, the true cause or nature of an outage is not known or recorded because crews concentrate on fixing the power system, rather than noting causes of outages in detail on the repair log. The inexact nature of this data collection presents difficulties for structural engineers because the degree to which the various system components can be redesigned or modified to result in prescribed structural damage more easily repaired than at present is not obvious. This policy based approach that analyzes the patterns of disruption is one tool that can be used to supplement the data from the repair logs.

ACKNOWLEDGEMENTS

Seattle City Light is gratefully acknowledged for allowing the writers to employ data for this analysis. Suggestions provided by Mr. Dave Albergine were extremely helpful. Funding from the National Science Foundation Grant Number CMS 0099638 is gratefully acknowledged. Useful suggestions were provided by Prof. Nojima of Gifu University. The survey data provided by Meszaros and Fiegener were obtained in a study funded by the Economic Development Administration, Department of Commerce, and the Pacific Earthquake Engineering Research Center (PEER).

REFERENCES

Anagnos, T. (1999) "Improvement of Fragilities for Electrical Substation Equipment," *Proc. of the Fifth US Conference on Lifeline Earthquake Engineering*, Ed. by W. Elliott and P. McDonough, New York: American Society of Civil Engineers, pp. 673-682.

Applied Technology Council (1991), "Seismic Vulnerability and Impact of Disruption of Lifelines in the Conterminous United States," *ATC-25*, Redwood City, California.

Arcview GIS software, http://www.esri.com/software/arcview/.

ASCE 7-02 (2002) *Minimum Design Loads for Buildings and Other Structures*, ASCE Publications.

Brown, R.E., S. Gupta, R.D. Christie, S.S. Venkata and R. Fletcher (1997a), "Automated Primary Distribution System Design: Reliability and Cost Optimization," *IEEE Trans. on Power Delivery*.

Brown, R.E., S. Gupta, R.D. Christie, S.S. Venkata and R. Fletcher (1997b), "Distribution System Reliability Assessment: Momentary Interruptions and Storms," *IEEE Trans. on Power Delivery*.

Brown, R.E., S. Gupta, R.D. Christie, S.S. Venkata and R. Fletcher (1997c), "Distribution System Reliability Assessment Using Hierarchical Markov Modeling," *IEEE Trans. on Power Delivery*.

Chang, S. (1998), "Direct Economic Impacts," *Engineering and Socioeconomic Impacts of Earthquakes,: An Analysis of Electricity Lifeline Disruptions in the New Madrid Area*, MCEER: University of Buffalo, Buffalo, NY, pp. 75-94.

Davidson, R.A., H. Liu, I.K. Sarpong, P. Sparks and D. V. Rosowsky (2003), *Natural Hazards Review*, Feb., pp. 36-45.

Department of Earth and Spaces Science at the University of Washington (www.ess.washington.edu/shake/).

Distribution System Reliability, 2002, Seattle City Light Distribution Branch Report, Jan. 3, 2002.

EIA (2001), http://www.eia.doe.gov/cneaf/electricity/st_profiles/washington/wa.html.

FEMA (2000), HAZUS-99, Service Release 2, www.fema.gov/hazus.

Hosmer, David W. and Lemeshow, Stanley, (2000), *Applied Logistic Regression*, Wiley Series in Probability and Statistics, 2nd edition, Wiley: New York, NY, 375pp.

Hwang, Howard H.M. and Jun-Rong Huo (1995), "Seismic Performance Evaluation of Substation Structures." In O'Rourke, Michael J. (ed.), *Lifeline Earthquake Engineering: Proceedings of the Fourth U.S. Conference; San Francisco, California, August 10-12, 1995.*, (Technical Council on Lifeline Earthquake Engineering Monograph No. 6) New York: American Society of Civil Engineers.

IEEE Guide for Electric Power Distribution Reliability Indices, IEEE Standard 1366, 2001 Edition.

Meszaros, Jacqueline and Mark Fiegener (2002) *Effects of the 2001 Nisqually Earthquake on Small Businesses in Washington State*, Report prepared for the Economic Development Administration, US Department of Commerce, October.

Nojima, N. and M. Sugito, (2002) "Empirical Estimation of Outage and Duration of Lifeline Disruption due to Earthquake Disaster," *Proc. of the US-China-Japan Workshop on Lifeline Systems*, October.

O'Rourke, T.D., S.S. Jeon, R.T. Eguchi and C.K. Huyck, (2001) "Advanced GIS for Loss Estimation and Rapid Post-Earthquake Assessment of Building Damage," *MCEER Research Progress and Accomplishments 2000-2001,* available from http://www.mceer.buffalo.edu/.

Petak, William J. (2003) "Earthquake Mitigation Implementation: A Sociotechnical Systems Approach," presented at the 55th Annual Meeting of the Earthquake Engineering Research Institute, Feb. 6th.

Reed, D.A. and Carolyn Cook (1999), "Multi-Hazard Analysis of Utility Lifelines," *Proc. of the 10th International Conference on Wind Engineering,* Copenhagen, Denmark, June.

Reed, D.A., J. Preuss, and J. Park (2002a), "Empirical Investigation of Utility Lifeline Restoration for Earthquake and Wind Hazards," *Proc. Of Hazards 2002*, Antalya, Turkey, Oct. 2-6.

Reed, D.A., J. Preuss and J. Park (2002b), "Distribution System Disruption and Recovery for Natural Hazards," *Proc. of Power Systems and Communications Infrastructures for the Future,* Beijing, PRC, Sept.

Schiff, A. (1998) "Electric Power," *Hyogoken-Nanbu (Kobe) Earthquake of January 17,1995 Lifeline Performance,* Ed. By Anshel Schiff, Technical Council on Lifeline Earthquake Engineering, Monograph No. 14, New York: American Society of Civil Engineers, Sept. 1998.

S-Plus statistical analysis software, http://www.spss.com.

Tanner, C.D. (1991), *Potential Intertidal Habitat Restoration Sites in the Duwamish River Estuary*, EPA 910/9-91-050. Final report prepared for the US EPA and Port of Seattle. Seattle, WA. 93 p. + appendices.

Wald, D.J., V. Quitoriano, T.H. Heaton, H. Kanamori (1999), "Relationships between Peak Ground Acceleration, Peak Ground Velocity and Modified Mercalli Intensity in California," *Earthquake Spectra*, Vol. 15, No. 3, pp. 557-564.

RISK CRITERIA ISSUES

Criteria for acceptable risk in the Netherlands

J.K. Vrijling[a], P.H.A.J.M. van Gelder[a] & S.J. Ouwerkerk[a]
[a] Delft University of Technology, P.O. Box 5048, 2600 GA Delft, the Netherlands;
E-mail: j.k.vrijling@ct.tudelft.nl, p.vangelder@ct.tudelft.nl,
sonjaouwerkerk@yahoo.com

Abstract
Risk criteria are reference levels that are set in order to protect people against natural and man-made hazards. In the Netherlands, discussion has risen about the current risk criteria for the field of external safety. Reason for the discussion can be found in the fact that risk has been customarily considered purely as the probability of the loss of life. Other aspects such as economical damage and the degree to which the exposure to the risk is voluntary are not taken into account. To judge risk in a wider context a set of rules for the evaluation of risk, which leads to technical advice in a question that has to be decided politically, is proposed.

1 Introduction

Like many countries, the history of the Netherlands is marked by numerous disasters. Our prehistoric ancestors were threatened by natural hazards like extreme weather, floods and wild animals. Since the industrial revolution man-made hazards such as industrial accidents, train derailments, tunnel fires and airplane crashes also disrupt society on a regular basis (Jonkman et al, 2003). One of the first signs of the man-made hazards of the industrial revolution in the Netherlands was the explosion of a powder tower in the centre of Delft in 1654, resulting in the destruction of two hundred houses and the deaths of about hundred citizens. But also today the inhabitants of the Netherlands are frequently startled by the occurrence of both natural and man-made hazards. For example in 1953, a large part of the Netherlands was flooded due to a severe northwestern storm and over 1800 people lost their lives. Almost 40 years later, in 1992, one of the most devastating man-made hazards occurred: an Israeli cargo plane crashed into a 12-story apartment block in the Amsterdam suburb of Bijlmer. At least 39 people (reported) on the ground and all 4 people aboard the aircraft were killed. In May 2000 the thinking about safety and risks in the Netherlands took a new direction. A disaster occurred in Enschede, a city in the east of the Netherlands, that nobody had thought possible. The explosion of a firework warehouse wiped out an entire residential area. Thousands of buildings were damaged and there were about 20 fatalities.

A risk-free society without risk is not possible and not desirable, as risky activities are an engine for economic growth, but in order to prevent that certain inhabitants are exposed to a disproportionately large risk, risk criteria are applied. The current risk criteria in the Netherlands are under discussion. On the one hand, risky activities on occasion do not comply with the risk criteria. An example is the expansion of the Dutch national airport Schiphol, where risk criteria have been adapted in order to facilitate growth of the airport. On the other hand, some activities satisfy the risk criteria, but are not allowed to take place. An example is the nuclear power plant at

Borssele. Many people want the plant to be closed down, disregarding the fact that the plant complies amply with both national and international norms. Apparently, no consensus exists about the acceptability of risk as laid out by De Hollander and Hanemaaijer (2003).

Today, large interest exists in the Netherlands in the responsible management of risks (Advisory Council for Transport, Public Works and Water Management and the Council for Housing, Spatial Planning and the Environment, 2003). This paper proposes a framework that can serve as a rational basis for technological design. Although its focus is on the Dutch situation the broad outline is generally applicable.

2   Risk policy in the Netherlands

In order to protect the inhabitants of the Netherlands against risk resulting from dangerous activities, risk criteria are set. Risk criteria are reference levels against which the results of a risk analysis should be assessed (Vrouwenvelder et al, 2001). The first risk criteria date back from 1810 when the emperor Napoleon issued a decree stating that a permit was needed to operate an industry (Ale, 2002). In the imperial decree, a distinction was made between activities that were allowed inside cities and activities that were only allowed at certain distances outside housing development. It was not until 1960 that probabilistic methods were introduced in the risk policy. Van Dantzig (1956) and Van Dantzig and Kriens (1960) used an economical optimization of the height of the flood defenses along the Dutch coast resulting in a minimum safety level of main sea dikes of 10^{-4}/yr. The external safety policy in its current form originated in the beginning of the 1980s, when it became clear that the use of LPG would increase considerably (Bottelberghs, 2000). In those years an evaluation system was developed that was based on quantitative assessment of risks and quantitative criteria for decisions on risk acceptability. Nowadays, risk assessment techniques are applied in policy and regulation for several fields, such as the use of airports or the transport of hazardous materials.

In the current Dutch risk policy the specified level of harm is considered from two points of view. One is the point of view of the individual, who decides to undertake an activity weighing the risks against the direct and indirect personal benefits. The individual risk for a point location around a hazardous activity is defined as the probability that an average unprotected person (hypothetically) permanently present at that point location, would get killed due to an accident at the hazardous activity (Bottelberghs, 2000). Individual risk depends on the geographic position and can be presented as iso-risk contours on a map by drawing lines that connect locations with the same level of risk. The iso-risk contours give information about the risk at a certain location, regardless whether people are present at that location or not. The second point of view is the one of the society, considering whether or not an activity is acceptable in terms of the risk-benefit trade off for the total population. The societal risk for a hazardous activity is defined as the probability that a group of more than N persons would get killed due to an accident at the hazardous activity area (Bottelberghs, 2000). Societal risk is characteristic for the hazardous activity in combination with the population density in the surroundings. It can be presented in the form of a probability mass function: an fN-curve. In an FN-curve, however, the probability of exceedance or cumulative frequency, $F(\geq N)$, of N or more fatalities per year is plotted, where $F(\geq N) = \Sigma f(N)$, summed from N to N_{max}. Figure 1(a,b) shows an example of both a probability mass function and a frequency of exceedance curve.

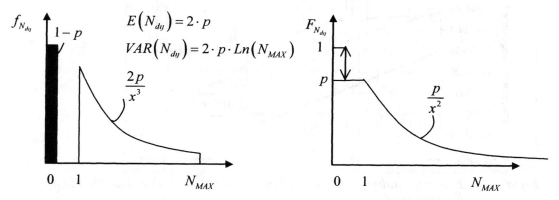

Figure 1 (a) probability mass function for the number of deaths by an inverse quadratic Pareto; (b) probability of exceedance curve for the number of deaths by an inverse quadratic Pareto.

For both risk criteria limits have been set (Ale and Piers, 2000). The present values of these limits for industrial activities are determined by the Ministry of Housing, Spatial Planning and the Environment (VROM). The individual risk limit is set to 10^{-6}/yr for new situations and 10^{-5}/yr for existing situations. These are limit values under the law, which means that they cannot be exceeded. The societal risk limit is set at $F=10^{-3}/N^2$, which serves as a guideline. In practice, not all activities comply with the current risk criteria. To illustrate the problems concerning the current risk criteria, they are applied below to a number of activities, namely the national airport Schiphol, LPG-stations and road safety.

Schiphol
At Schiphol national airport about 200,000 planes leave and arrive every year bringing the total number of movements to 400,000 per year. As Schiphol is surrounded by inhabited areas this leads to the exposure of a considerable amount of people to risks above the individual risk level (Ale and Piers, 2000). In 2001, almost 4100 people were exposed to risks larger than the VROM-limit of $1 \cdot 10^{-6}$/yr and about 50 people were even exposed to risks larger than $1 \cdot 10^{-5}$/yr. The societal risk criterion is exceeded as well. Figure 2 shows that the probability of an accident with 100 or more fatalities is equal to once in 70,000 years, while the limiting value according to VROM requires once in 1,000,000 years.

In 2003, the government adapted a new policy in order to control the further growth of the risk. The policy states that it is not allowed to build within the 10^{-5}-contours and that the current safety situation may not deteriorate. In 2010, no inhabitants will be allowed within the $5 \cdot 10^{-5}$-contours. Apparently, the economic importance of Schiphol allows a larger risk for Schiphol than for other industrial activities.

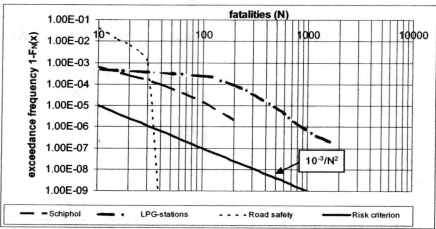

Figure 2   FN-curve (adapted from: National Institute for Public Health and the Environment, 2004).

LPG filling stations

About 2200 LPG filling stations are situated in the Netherlands. A part of these stations are located in inhabited areas, resulting in the exposure of 29,000 people to risks larger than 10^{-6}/yr of which more than 900 people were exposed to risks larger than 10^{-5}/yr in 2003. The presence of LPG-stations leads also to a large exceedance of the societal risk criterion (see figure 2): the probability of an accident with more than 100 fatalities is once per 5000 year.

The Dutch government has announced a three-year restructuring project involving approximately 200 LPG filling stations (Ministry of Housing, Spatial Planning and the Environment, 2004). EUR 15 million is earmarked to finance the removal of all LPG filling stations with facilities within the 10^{-5}-contours.

Road safety

In the Netherlands almost 1100 people die in the traffic every year. Given a population of 16 million people and assuming that every inhabitant of the Netherlands is exposed to risks resulting from traffic, this implies an individual risk of $1.4 \cdot 10^{-4}$ for each citizen, which exceeds amply the individual risk criterion of $1 \cdot 10^{-6}$ /yr. In view of societal risk, it is a smaller problem: most of the traffic accidents cost less than 10 fatalities. However, bus accidents can cost 20 or 30 lives. As can be seen in figure 2, the FN-characteristics of road safety result in a steep line.

The government wants the number of fatalities to be reduced to 980 in 2006 (Ministry of Transport, Public Works and Water Management, 2004). This would result in a reduction of the individual risk from $1.4 \cdot 10^{-4}$ to $1.6 \cdot 10^{-4}$ per year, which amounts to a large exceedance of the individual risk criterion set by VROM. Apparently, a daily activity with personal benefit and a relative high degree of voluntariness is more acceptable to society.

3   A framework for the acceptability of risk

In the Dutch risk policy, risk is narrowed down to the probability of the loss of life. However, the concept 'risk' involves many dimensions: it is characterized by a mixture of both technical and non-technical aspects. Technical scientists determine risk by measurement and calculation. In this approach, risk is frequently defined as 'the product of the probability of an event and its (monetary) consequences'. Probabilities and consequences of an event are quantified and combined in a risk number, which is used as the base for decision-making. Non-technical scientists attribute much value to the perception of risk. Risk perception deals with the judgments people make, when they are asked to characterize and evaluate hazardous activities and technologies. Vlek (1996) compiled a list of basic dimensions underlying perceived riskiness (table I).

Table I   **basic dimensions underlying perceived riskiness (adapted from Vlek (1996)).**

1.	Potential degree of harm or fatality
2.	Physical extent of damage (area affected)
3.	Social extent of damage (number of people involved)
4.	Time distribution of damage (immediate and/or delayed effects)
5.	Probability of undesired consequence
6.	Controllability (by self or trusted expert) of consequences
7.	Experience with, familiarity, imaginability, of consequences
8.	Voluntariness of exposure
9.	Clarity, importance of expected benefits
10.	Social distribution of risks and benefits
11.	Harmful intentionality

It is assumed that acceptable risk should be seen in a wider context. Since it cannot be judged separately from other aspects of the activity. The acceptance can only be understood in a cost-benefit framework in the widest sense. Therefore, a set of rules is proposed, using two points of view: a personal and a societal. Personal gain, national gain, capital outlays, running costs, damage to the environment, and the risk play a part in the weighing process. As this complicated process cannot be adequately modeled, two crude approximations are proposed in this paper. The first is to accept the pattern of the accident statistics as the outcome of the cost-benefit weighing. The second is a risk-oriented technical cost-benefit model that expresses all consequences of failure in monetary terms.

Personally acceptable level of risk
The smallest-scale component of the social acceptance of risk is the personal cost-benefit assessment by the individual. Since attempts to model this appraisal procedure quantitatively are not feasible, it is proposed to look at the pattern of preferences revealed in the accident statistics. The fact that the actual personal risk levels connected to various activities show statistical stability over the years and are approximately equal for the Western countries indicates a consistent pattern of preferences. The probability of losing one's life in normal daily activities such as driving a car or working in a factory appears to be one or two orders of magnitude lower than the overall probability of dying. Only a purely voluntary activity such as

mountaineering entails a higher risk (figure 3). This observation of public tolerance of 1000 times greater risks from voluntary than from involuntary activities with the same benefit was already made by Starr (1969).[1]

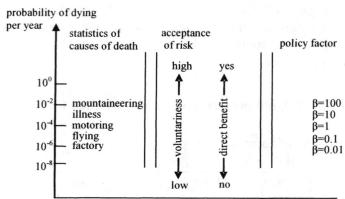

Figure 3   personal risk in Western countries, deduced from the statistics of causes of death and the number of participants per activity.

In view of the consistency and the stability, apart from a slightly downward trend due to technical progress, of the death risks presented, it seems permissible to use them as a basis for decisions with regard to the personally acceptable probability of failure in the following way:

$$P_{fi} = \frac{\beta_i \cdot 10^{-4}}{P_{d|fi}} \tag{1}$$

where P_{fi} is the yearly probability of dying and $P_{d|fi}$ denotes the probability of being killed in the event of an accident. In this expression the policy factor β_i varies with the degree of voluntariness with which an activity i is undertaken and with the benefit perceived. It ranges from 100, in the case of complete freedom of choice like mountaineering (P_{fi},= 0.1 = 100*10^{-4}/10^{-1}) to 0.01 in the case of an imposed risk without any perceived direct benefit.

A proposal for the choice of the value of the policy factor β_i as a function of voluntariness and benefit is given in table II. It should be noted that a β_i-value has to be chosen for each threatened group, that differs in relation to the activity. For instance, the pilots, passengers and people living under the flight paths each have a specific relation to air travel and consequently different visions on the acceptability of a certain level of risk.

---

[1] It is noted that people tend to reject risks when asked directly (Fischhoff, 1990). However, in their more anonymous role as a citizen of the society they effectively accept it.

Table II   **the value of the policy factor β_i as a function of voluntariness and benefit.**

β_i	Voluntariness	Direct benefit	Example
100	Completely voluntary	Direct benefit	Mountaineering
10	Voluntary	Direct benefit	Motor biking
1.0	Neutral	Direct benefit	Car driving
0.1	Involuntary	Some benefit	Factory
0.01	Involuntary	No benefit	LPG-station

Societal acceptable level of risk

The base of the framework with respect to societal risk in the Netherlands, is an evaluation of risks due to a certain activity on a national level. The risk on a national level is the aggregate of the risks of local installations or activities. Without mentioning it specifically, the risk criteria as developed by VROM in the Netherlands are meant to support a systematic appraisal by the local authorities of a single installation or activity.

If a risk criterion is thus defined on a local level the height of the national risk criterion is determined by the number of locations, where the activity takes place and by the probability mass function of the consequences of an accident. The acceptability of the resulting national norm has to be assessed separately, as it was not intentionally formulated.

It seems preferable to start with a risk criterion on a national level and to evaluate the acceptable local risk level, in view of the actual number of installations, the cost-benefit aspects of the activity and the general progress in safety, in an iterative process with, say, a 10-year cycle (figure 4).

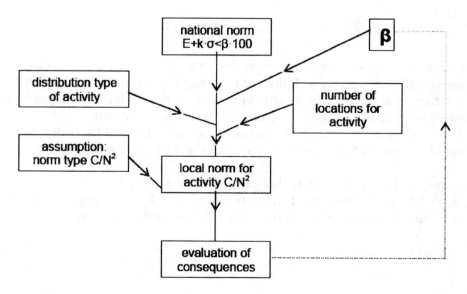

Figure 4   flowchart for risk management

Nationally acceptable level of risk

The determination of the socially acceptable level of risk starts from the assumption that the accident statistics reflect the result of a social process of cost-benefit appraisal. If these statistics reveal the preferences, a standard can be derived from them. It can be shown that the very low probabilities of a fatal accident, which appear socially acceptable, are perceptible using the concept of the circle of acquaintances as an instrument of observation. The recurrence time of an accident, claiming the life of an acquaintance from the circle, is of the order of magnitude of a human life span. To establish a norm for the acceptable level of risk for engineering structures it is more realistic to base oneself on the probability of a death due to a non-voluntary activity in the factory, on board a ship, at sea, etc., which is approximately equal to 1.4×10^{-5}/year, than on the number of casualties in the car traffic, which seems on the verge of acceptance. If this observation-based frequency is adopted as the norm for assessing the safety of activity i, then after rearranging the expression, and adopting a rather arbitrary distribution over some 20 categories of activities, each claiming an equal number of lives per year, the following norm is obtained for an activity i with N_{pi} participants in the Netherlands:

$$P_{fi} \cdot N_{pi} \cdot P_{d|fi} < \beta_i \cdot 100 \tag{2}$$

Note that the factor 100 is country-specific and based on: the value of the minimum death rate of the population, the ratio of the involuntary accident death rate (exclusive diseases) with the minimum death rate, the number of hazardous activities in a country (in average 20 sectors) and the size of the population of the country. Comparing this multiplication factor for the Netherlands (MF_{NL}) with the factor for South Africa (MF_{SA}), the factor for South Africa appears to be about 7 times higher than for the Netherlands:

$$MF_{NL} = \frac{10^{-4} \cdot 15 \cdot 10^6}{7 \cdot 20} \approx 100$$

$$MF_{SA} = \frac{10^{-3} \cdot 45 \cdot 10^6}{3 \cdot 20} \approx 750$$

Formula (2) states that an activity is permissible as long as it is expected to claim fewer than $\beta_i \cdot 100$ deaths per year. It does not account for risk aversion, which will certainly influence acceptance by a community or a society. Relatively frequent small accidents are more easily accepted than one single rare accident with large consequences, although the expected number of casualties is equal for both cases.[2] The standard deviation of the number of casualties will reflect this difference.

Risk aversion can be represented mathematically by increasing the mathematical expectation of the total number of deaths per year, $E(N_{di})$, by an appropriate multiple k of the standard deviation before the situation is tested against the norm:

$$E(N_{di}) + k \cdot \sigma(N_{di}) < \beta_i \cdot 100 \tag{3}$$

---

[2] It is noted that Slovic et al. (1994) shed doubt on this assumption, but here risk aversion is adopted.

where k = risk aversion index. To determine the mathematical expectation and the standard deviation of the total number of deaths occurring annually in the context of activity *i*, it is necessary to take into account the number of independent places N_{Ai} where the activity under consideration is carried out.

The norm with k = 3 is tested for several activities in the Netherlands by Vrijling et al. (1995). The agreement between the norm for reasonable values of N_{Ai} and $0.01 < \beta_i < 100$ and the actual risks accepted in practice seems to support the model.

Locally acceptable level of risk
The translation of the nationally acceptable level of risk to a risk criterion for one single installation or location where an activity takes place depends on the distribution type of the number of casualties for accidents of the activity under consideration. In order to relate the new local risk criterion to the present one proposed by VROM ($C_i = 10^{-3}$), a societal risk criterion of the following type is preferred:

$$1 - F_{N_{di}}(x) < \frac{C_i}{x^2} \text{ for all } x \geq 10 \tag{4}$$

where x is the number of deaths.

Assuming a Bernoulli distribution (a distribution that limits the outcomes to zero or N fatalities) for the number of casualties at each of N_{Ai} independent locations, the expected value and the standard deviation of the casualties at national level are:

$$E(N_{di}) = N_{Ai} \cdot p_{fi} \cdot N_{dij|f}$$

$$\sigma(N_{di})^2 = N_{Ai} \cdot p_{fi} \cdot (1 - p_{fi}) \cdot N_{dij|f}^2 \approx N_{Ai} \cdot p_{fi} \cdot N_{dij|f}^2 \tag{5}$$

where N_{Ai} is the number of independent locations, p_{fi} and $N_{dij|f}$ are the probability of failure at a location and the number of fatalities given failure, respectively.

If the Bernoulli distribution of the number of casualties at each location complies with criterion (4), it follows that for a location $E(N_{dij}) \leq C_i/N$ and $\sigma(N_{dij}) \leq C_i$. Substituting these values in equation (5) and subsequently in the national criterion (3), and solving the resulting quadratic equation in p_{fi}, gives for the value of C_i:

$$C_i = \left[\frac{-k \cdot \sqrt{N_{Ai}} + \sqrt{k^2 \cdot N_{Ai} + 4\frac{N_{Ai}}{N} \cdot \beta_i \cdot 100}}{2\frac{N_{Ai}}{N}} \right]^2 \tag{6}$$

If the expected value of the number of deaths is much smaller than its standard deviation, which is often true for the rare calamities studied here, the previous result reduces to:

$$C_i \approx \left[\frac{\beta_i \cdot 100}{k \cdot \sqrt{N_{Ai}}} \right]^2 \tag{7}$$

Similar results are obtained if the conditional probability mass function (p.m.f.) of the number of deaths is geometric instead of Dirac. (Dirac is a conditional p.m.f. that limits the outcomes to exactly N fatalities.) The national societal acceptable risk criterion leads to a local acceptable risk criterion of the VROM-type, which is inversely proportional to the number of independent places N_A and the square of the policy factor β_i:

$$1 - F_{N_{d_t}}(x) \le \frac{C_i}{x^2} \text{ for all } x \ge 10, \text{ where } C_i = \left[\frac{\beta_i \cdot 100}{k \cdot \sqrt{N_{A_i}}}\right]^2 \tag{8}$$

Surprisingly, it can be mathematically proven that the national rule given by eqn (3) is more strict than the norm defined by eqn (8). Theoretically eqn (8) allows an infinite standard deviation, while the national rule limits this. In cases where the expected values are not negligible, one is advised to work from the basic formulae:

$$E(N_{d_t}) = N_{Ai} \cdot p_{fi} \cdot E(N_{dij|f})$$
$$\sigma(N_{d_t})^2 = \left[N_{Ai} \cdot p_{fi} \cdot (1 - p_{fi}) \cdot \left(E(N_{dij|f})^2 + \sigma(N_{dij|f})^2\right)\right] \tag{9}$$

Substitution of these in the national rule given by eqn (3) will result in a limitation on p_{fi} for each location of activity i.

The VROM-rule is a special case of this general rule for acceptable risk. For instance, for chemical installations in the Netherlands with $C_i = 10^{-3}$, $N_A = 1000$ (the approximate number of chemical installations) and $k = 3$, it follows that $\beta = 0.03$ which is according to figure 3 not unreasonable for an involuntarily imposed risk.

Car traffic forms another interesting example, because the number of independent installations $N_A = 4 \cdot 10^6$ is very large and the victims are passengers/users. If the number of people in the car is assumed to be 2 and the probability to die in a crash is $p_{d|f} = 0.1$, then the conditional expectation and the standard deviation of the number of deaths per car are equal to 0.2 and 0.42 respectively. Using the general formulae mentioned, the expected value and the standard deviation at national level are calculated:

$$E(N_{d_t}) = N_{A_i} \cdot p_{fij} \cdot E(N_{dij|f})$$
$$\sigma(N_{d_t}) = \left[N_{A_i} \cdot p_{fij} \cdot \left((1 - p_{fij}) \cdot E(N_{dij|f})^2 + \sigma(N_{d|f})^2\right)\right]^{1/2} \tag{10}$$

If $\beta = 1.0$ is adopted the acceptable probability of a car accident is $0.9 \cdot 10^{-4}$ per year per individual. The expected total number of casualties amounts to 72 per year with a standard deviation of 8.8. A choice of $\beta_i = 10$ leads to an increase of the acceptable probability of an accident to $1.1 \cdot 10^{-3}$ per car per year. The expected total number of casualties amounts to 972 per year with a standard deviation of 30.8. This is more in line with the actual situation, where the traffic claims approximately 1100 lives per year.

Economically acceptable level of risk

The problem of the acceptable level of risk can also be formulated as an economic decision problem. According to the method of economic optimisation, the total costs in a system (C_{tot}) are determined by the sum of the expenditure for a safer system (I) and the expected value of the economic damage. In the optimal economic situation the total costs in the system are minimised:

$$\min(Q) = \min\left(I(P_f) + PV(P_f \cdot S)\right) \tag{11}$$

where Q=total cost, PV=present value operator and S=total damage in case of failure.

If, despite ethical objections, the value of a human life is rated at s, the amount of damage is increased to:

$$P_{d|fi} \cdot N_{pi} \cdot s + S \tag{12}$$

where N_{pi} = number of participants in activity i. This extension makes the optimal failure probability a decreasing function of the expected number of deaths. The valuation of human life is chosen as the present value of the net national product (NNP) per inhabitant.[4] The advantage of taking the possible loss of lives into account in economic terms is that the safety measures are affordable in the context of the national income. Risk aversion may also be included in the economic approach as shown by Van Gelder and Vrijling (1997).

4 Conclusions

In the Dutch risk policy two points of view are considered: the point of view of the individual, who decides to undertake an activity weighing the risks against the direct and indirect personal benefits and the point of view of the society, considering if an activity is acceptable in terms of the risk-benefit trade-off for the total population. However, many activities, however, such as Schiphol national airport, LPG-stations and road safety amply exceed both individual and societal risk criteria. As the current risk criteria in the Netherlands are under discussion, a set of rules has been proposed that can indicate the acceptable probability of failure of technical systems over a wide range of economic activities and for categories of people with different relations to the system (personnel, users and third parties). Two points of view are discerned: a personal and a societal point of view. These points of view have resulted in three risk criteria: a personally acceptable level of risk, a socially acceptable level of risk, and an economically acceptable level of risk. The most stringent of the three criteria should be adopted as a basis for the technical advice to the political decision process. However, all information about the risk assessment should be available in the political process. A broad societal dialogue both at national and international level is needed in order to establish more clearly which factors play a role and which quantitative values (of risk) are tolerable.

---

[3] It should be noted that risk aversion and uncertainties are not included in the above economic criterion. In Slijkhuis et al (1997) uncertainties are taken into account for the calculation of the height of a sea dike, and Van Gelder and Vrijling (1997) have investigated the influence of risk aversion on the economic optimisation.

[4] Many wish to use an estimate based on the CSX-value (the cost of saving an extra life), which seems not correct (Vrijling and Van Gelder, 2000). Here, the more stable relation to NNP is preferred.

Finally, it should be realized that the philosophy and the techniques set out above are just means to reach a goal. One should not lose sight of the goal of managed safety, when dealing with the tools that serves instruments to measure aspects of the entire situation.

References

Advisory Council for Transport, Public Works and Water Management and the Council for Housing, Spatial Planning and the Environment, 2003. *Verantwoorde risico's, veilige ruimte (in Dutch)*. Den Haag.

Ale, B.J.M., 2002. *Risk assessment practices in The Netherlands*. Safety   Science 40, 105-126.

Ale, B.J.M. and M. Piers, 2000. *The assessment and management of third party risk around a major airport*. Journal of Hazardous Materials 71, 1-   6.

Bottelberghs, P.H., 2000. *Risk analysis and safety policy developments in the Netherlands*. Journal of Hazardous Materials 71, 59-84.

De Hollander, A.E.M. and A.H. Hanemaaijer, 2003. *Coping rationally with risks (in Dutch)*. RIVM report 251701047.

Fischhoff, B., 1990. *Psychology and public policy*. American Psychologist, 45(5), 647-653.

Jonkman, S.N., P.H.A.J.M. van Gelder and J.K. Vrijling, 2003. *An overview of quantitative risk measures for loss of life and economic damage*. Journal of Hazardous Materials A99, 1-30.

Ministry of Housing, Spatial Planning and the Environment, 2004. *Ministry of Housing, Spatial Planning and the Environment. http://www.vrom.nl*

Ministry of Transport, Public Works and Water Management, 2004. *Ministry of Transport, Public Works and Water Management.* http://www.verkeerenwaterstaat.nl

National Institute for Public Health and the Environment, 2004. *Dutch Environmental Data Compendium (in Dutch).* http://www.rivm.nl/milieuennatuurcompendium/nl

Slijkhuis, K.A.H., P.H.A.J.M. van Gelder and J.K. Vrijling, 1997. *Optimal dike height under statistical- construction- and damage uncertainty.*   Structural Safety and Reliability, Vol.7, pp. 1137-1140.

Slovic, P., S. Lichtenstein and B. Fischhoff, 1994. *Modeling the societal   impact of fatal accidents*. Management Science, 30(4), 464-474.

Starr, C., 1969. *Social benefit versus technological risk.* Science,165, 1232-1283.

Van Dantzig, V.D., 1956. *Economic decision problems for flood prevention.* Econometrica, 24, 276-287.

Van Dantzig, V.D. & J. Kriens, 1960. *The economic decision problem of safeguarding the Netherlands against floods. Report of Delta Commission, Part 3, Section II.2 (in Dutch).* The Hague.

Van Gelder, P.H.A.J.M. and J.K. Vrijling, 1997. *Risk-Averse Reliability-Based Optimization of Sea-Defences.* Risk Based Decision Making in Water Resources, Vol. 8, pp. 61-76.

Vlek, C.A.J., 1996. A multi-level, multi-stage and multi-attribute perspective on risk assessment, decision-making and risk control. Risk decision and Policy 1 (1), 9-31.

Vrijling, J.K., W. van Hengel and R. J. Houben, 1995. *A framework for risk evaluation.* Journal of Hazardous Materials. Vol. 43, Issue 3, 245-261.

Vrijling, J.K. and P.H.A.J.M. van Gelder, 2000. *An Analysis of the Valuation of a Human Life,* in: PRECAUTION, Volume 1, pp.197-200, Eds. M.P. Cottam, D.W. Harvey, R.P. Pape, and J.Tait, May 14-17 2000 Edinburgh, Scotland, UK.

Vrouwenvelder, A., B.M. Holicky, C.P. Tanner, D.R.Lovergrove, E.G. Canisius, 2001. Risk assessment and risk communication in Civil Engineering. CIB Report: Publication 259; ISBN: 90-6363-026-3.

SYSTEM MANAGEMENT ISSUES

LANDSLIDE RISK ASSESSMENT AND REMEDIATION

Mihail Popescu, Illinois Institute of Technology, Chicago
Manoochehr Zoghi, University of Dayton

1. Foreword

Landslides are frequently responsible for considerable losses of both money and lives. The severity of the landslide problem intensifies with increased urban development and change in land use. In view of this consideration, it is not surprising that landslides are rapidly becoming the focus of major scientific research, engineering study and practice, and land-use policy throughout the world (Popescu et al., 2001). International cooperation among various individuals concerned with the fields of geology, geomorphology, and soil and rock mechanics has recently contributed to improvement of our understanding of landslides, notably in the framework of the United Nations International Decade for Natural Disaster Reduction (1990-2000). Evidently, this provided the environment for establishing the International Geotechnical Societies' UNESCO Working Party on World Landslide Inventory (WP/WLI) which in 1994 became the International Union of Geological Sciences (IUGS) Working Group on Landslides (WG/L).

Landslides and related slope instability phenomena plague many parts of the world. Japan leads other nations in landslide severity with projected combined direct and indirect losses of $4 billion annually (Schuster, 1996). United States, Italy, and India follow Japan, with an estimated annual cost ranging between $1 billion to $2 billion. Landslide disasters are also common in developing countries and monetary losses sometimes equal or exceed the gross national product of these countries.

The paramount importance of landslide hazard management is by and large recognized. Herein lies the guiding principle of the current chapter; i.e., to describe landslide hazards and methods to mitigate the associated risks in an appropriate and effective manner.

2. Risk Management Process

The risk management process comprises two components: risk assessment and risk treatment. Landslide and slope engineering have always involved some form of risk management, although it was seldom formally recognized as such. This informal type of risk management was essentially the exercise of engineering judgment by experienced engineers and geologists.

Figure 1 shows the process of landslide risk management a in flow-chart form (Australian Geomechanics Society - AGS, 2000). In simple form, the process involves answering the following questions:

· What might happen?
· How likely is it?
· What damage or injury may result?
· How important is it?
· What can be done about it?

There is a clear distinction between hazard, risk and probability. Fell (1994) has defined the "hazard" as a condition with the potential for causing an undesirable consequence. The description of landslide hazard should include the location, volume (or area), classification, and velocity of potential landslides and any resulting detached material, and the probability of their occurrence within a given period of time.

Risk is a measure of the probability and severity of an adverse effect to health, property, or the environment. Risk is often estimated by the product of probability and consequences. However, a more general interpretation of risk involves a comparison of the probability and consequences in a non-product form.

Probability is the likelihood of a specific outcome, measured by the ratio of specific outcomes to the total number of possible outcomes. Probability is expressed as a number between 0 and 1, with 0 indicating an impossible outcome, and 1 indicating that an outcome is certain.

The intent of a landslide hazard assessment is to identify a region's susceptibility to landslides and their consequences based on a few key or significant physical attributes comprising the previous landslide activities, bedrock features, slope geometry, and hydrologic characteristics. In a development program (planning process) concerning a landslide-prone area, one needs to determine the acceptable risk. It is indispensable to recognize the vulnerability and degrees of risk involved and to instigate a systematic approach in avoiding, controlling, or mitigating existing and future landslide hazards in the planning process. Accordingly, either a planner should avoid the landslide-susceptible areas if it is deemed appropriate, or else he or she needs to implement strategies to reduce risk.

3. Landslide Hazard Identification

Landslide hazard identification requires an understanding of the slope processes and the relationship of those processes to geomorphology, geology, hydrogeology, climate and vegetation. From this understanding it will be possible to:

♦ Classify the types of potential landsliding - the classification system proposed by Varnes (1978) as modified by Cruden and Varnes (1996) constitutes a suitable system. It should be recognized that a site may be affected by more than one type of landslide hazard. For example, deep-seated landslides occur at the site, whereas, rockfall and debris flows will initiate from above the site.

♦ Assess the physical extent of each potential landslide being considered, including the location, areal extent, and volume involved.

♦ Assess the likely causal factor(s), the physical characteristics of the materials involved, and the slide mechanics.

♦ Estimate the resulting anticipated travel distance and velocity of movement.

♦ Address the possibility of fast-acting processes, such as flows and falls, from which it is more difficult to escape.

Methods that may be used to identify hazards include geomorphological mapping, gathering of historic information on landslides in similar topography, geology and climate, (e.g. from maintenance records, aerial photographs, newspapers, review of analysis of stability etc). Some forms of geologic and geomorphic mapping are considered to be an integrated component of the fieldwork stage when assessing natural landslides, which requires understanding the site while inspecting it.

3.1 Landslide Classification

The UNESCO Working Party's definition of a landslide is "the movement of a mass of rock, earth or debris down a slope" (Cruden, 1991, 1997) and recognizes that the phenomena described as landslides are not limited either to the land or to sliding; the word has a much more extensive meaning than its component parts suggest. An idealized diagram depicting the nomenclature of various features of a complex earth slide – earth flow is shown in Figure 2 (Cruden and Varnes, 1996).

As there is a wide spectrum of landslide types, those identified should be classified as far as feasible. The criteria used by the WP/WLI in classification of landslides follow Varnes (1978) in emphasizing the type of movement and type of material. The divisions of materials are: rock, debris and earth. Rock is characterized as an intact hard or firm mass in its natural place prior to the initial movement, whereas soil is referred to as an aggregate of unconsolidated solid particles, either transported or derived in-place via the weathering processes. The latter is further divided into earth in which 80% or more of the particles are smaller than 2 mm in size; and debris whereby 20% to 80% of the solid particles are larger than 2 mm.

Movements are divided into five types (Figure 3): falls, flows, slides, spreads and topples. In reality there is a continuum of mass movements from falls through slides to flows. In many instances it is difficult to determine whether masses of material have fallen or slid, and similarly there are a number of instances in which material has both slid and flowed. Very large falls can result in various types of flow involving fluidization with either water or air. The Department of Environment (DOE, 1994) recognized the existence of complex landslides in which ground displacement is achieved by more than one type of mass movement and emphasized that this should not be confused with landslide complex, i.e., an area of instability within which many different types of mass movement occur. Cruden and Varnes (1996) suggested that landslide complexity can be indicated by combining the five basic types of movement and the three divisions of materials. If the type of movement changes with the progress of movement, then the material should be described at the beginning of each successive movement. For example, a rock fall that has been followed by flow of the displaced material can be described as a rock fall–debris flow (Cruden and Varnes, 1996).

The landslide designation can become more elaborate as more information about the movement becomes available. Adjectives can be added in front of the noun string defining the type of landslide to build up the description of the movement. The adjective magnitude refers to the volume of displaced material involved in a landslide hazard, whereas the intensity renders a collection of physical parameters that describe the destruction or destructive potential of a landslide hazard. The qualitative expression of the former is small, medium, or large, whereas the latter is expressed qualitatively as slow, moderate, or fast as in downslope velocity of a debris flow. A landslide is known to be *active* when it is presently moving. An *inactive* landslide is one that last moved more than one annual cycle of seasons ago (Cruden and Varnes, 1996). Inactive landslides are further categorized into *dormant* if the causes of movement are apparent, and abandoned, if the triggering action is no longer present.

3.2 Landslide Causal Factors

"The processes involved in slope movements comprise a continuous series of events from cause to effect" (Varnes, 1978). It is of primary importance to recognize the conditions that caused the slope to become unstable and the processes that triggered the movement, when assessing landslide hazard for a particular site. Only an accurate diagnosis makes it possible to properly understand the landslide mechanisms and thence to propose effective treatment measures.

In every slope there are forces which tend to promote downslope movement and opposing forces which tend to resist movement. A general definition of the factor of safety of a slope results from comparing the downslope shear stress with the shear strength of the soil, along an assumed or known rupture surface. Starting from this general definition, Terzaghi (1950) divided landslide causes into external causes, which result in an increase of the shearing stress (e.g., geometrical changes, unloading the slope toe, loading the slope crest, shocks and vibrations, drawdown, changes in water regime) and internal causes that result in a decrease of the shearing resistance (e.g., progressive failure, weathering, seepage erosion). However, Varnes

(1978) pointed out that there are a number of external or internal causes which may be operating either to reduce the shearing resistance or to increase the shearing stress. There are also causes that simultaneously affect both terms of the factor-of-safety ratio.

The great variety of slope movements reflects the diversity of conditions that cause the slope to become unstable and the processes that trigger the movement. It is more appropriate to discuss causal factors (including both "conditions" and "processes") than "causes" per se alone. Thus ground conditions (weak strength, sensitive fabric, degree of weathering and fracturing) are influential criteria but are not causes (Popescu, 1996). They are part of the conditions necessary for an unstable slope to develop, to which must be added the environmental criteria of stress, pore-water pressure and temperature. It does not matter if the ground is weak as such–failure will only occur as a result if there is an effective causal process that acts as well. Such causal processes may be natural or anthropogenic, but effectively change the static ground conditions sufficiently to cause the slope system to fail, i.e. to adversely change the state of stability (Popescu, 1984).

Seldom, if ever, can a landslide be attributed to a single causal factor. The process leading to the development of a slide has its beginning with the formation of the rock itself, when its basic properties are determined and includes all the subsequent events of crustal movement, erosion and weathering (Varnes, 1978). The computed value of the factor of safety is a clear and simple distinction between stable and unstable slopes. However, from the physical point of view, it is better to visualize slopes existing in one of the following three stages: stable, marginally stable, and actively unstable (Crozier, 1986). Stable slopes are those where the margin of stability is sufficiently high to withstand all destabilizing forces. Marginally stable slopes are those that will fail at some time in response to the destabilizing forces having attained a certain level of activity. Finally, actively unstable slopes are those in which destabilizing forces produce continuous or intermittent movement.

The three stability stages must be seen to be part of a continuum, with the probability of failure being minute at the stable end of the spectrum, but increasing through the marginally stable range to reach certainty in the actively unstable stage. The three stability stages provide a useful framework for understanding the causal factors of landslides and classifying them into two groups on the basis of their function:

1. Preparatory causal factors which make the slope susceptible to movement without actually initiating it, and thereby tending to place the slope in a marginally stable state.

2. Triggering causal factors which initiate movement. The causal factors shift the slope from a marginally stable to an actively unstable state.

A particular causal factor may inflict either or both functions, depending on its degree of activity and the margin of stability. Although it may be possible to identify a single triggering process, an explanation of ultimate causes of a landslide invariably involves a number of preparatory conditions and processes. Based on their temporal variability, the destabilizing processes may be grouped into slow–changing (e.g. weathering, erosion) and fast–changing processes (e.g., earthquakes, reservoir drawdown). In the search for landslide causes, attention is often focused on those processes within the slope system that provoke the greatest rate of change. Although slow changes act over a long period of time to reduce the resistance/shear stress ratio, often a fast change can be identified as having triggered movement.

Because the assessment of landslide causes is complex and landslides are not always investigated in great detail, UNESCO World Party on World Landslide Inventory (1994) adopted a simple classification system of landslide causal factors. The operational approach to classification of landslide causal factors, proposed by these systems, is intended to cover the majority of landslides. It involves consideration of the available data from simple site investigation and information furnished by other site observations. Landslide causal factors are grouped according to their effect (preparatory or triggering) and their origin (ground conditions and geomorphological, physical or man-made processes). Ground conditions may not have a triggering function, while any ground condition or process may have a preparatory function.

Ground conditions or the material and mass characteristics of the ground, can be mapped on the surface of the landslide and the surrounding ground and explored in the subsurface by drilling, trenching and adits. Mechanical characteristics can be determined by testing. Geomorphic processes, or changes in the morphology of the ground, can be documented by pre-existing maps, aerial photographs, surveys of the landslide, or careful observation over time by the local population. Physical processes concern the environment and can be documented at the site by instrumentation, such as rainfall gauges, seismographs or piezometers. Careful local observations of water wells or damage from earthquakes may be acceptable substitutes. Variations in mechanical properties with distance from the surface may, in some circumstances, indicate changes of these properties with time. Man-made processes can be documented by site observations and from construction or excavation records at the site. Separate identification of man-made and natural landslides is useful for both administrative and theoretical reasons.

4. Regional Landslide Hazard Assessment

A landslide risk assessment, whether regional or local, has to be preceded by a corresponding landslide hazard assessment, from which it is then derived. The ordinary landslide hazard assessment techniques include deterministic, statistical, and heuristic (Barredo et al., 2000). Deterministic approaches are based on stability models and are utilized to map landslide hazards at large scales, typically for construction purposes (Einstein, et al., 2001). These models, however, necessitate detailed geotechnical and groundwater field data, which may not be always available.

The statistical approach, bivariate or multivariate, requires a large database, obtained from combination of factors that have initiated landsliding in the past. This model is generally useful for prediction of future landslides at medium scale and it is less suitable for reactivated slides.

The heuristic technique, also known as the knowledge-driven approach, is based on the analyst's expertise in identifying the type and degree of hazard for a designated area based on direct or indirect mapping. The direct method is accomplished either by directly mapping the degree of hazard in the field or by recording a geomorphic map. The indirect method utilizes an integration technique by combining several parameter maps based on qualitative weighing values assigned to each class of parameter map. This procedure is facilitated by compilation of various parameter layers via GIS for establishing the hazard assessment.

Before embarking on a regional landslide hazard assessment, the following preparatory steps are to be taken (Hutchinson, 2001):

1. Identify the user and purpose of the proposed assessment. Involve the user in all phases of the program.

2. Define the area to be mapped and decide the appropriate scale of mapping. This may range from 1: 100,000 or smaller to 1:5,000 or larger.

3. Obtain, or prepare, a good topographic base map of the area, preferably contoured.

4. Construct a detailed database of the geology (solid and superficial), geomorphology, hydrogeology, pedology, meteorology, mining and other human interference, history, and all other relevant factors within the area, and of all known mass movements including all published work, newspaper articles and the results of interviewing the local population.

5. Obtain all available air photo cover, satellite imagery and ground photography of the area. Photography of various dates can be particularly valuable, both because of what can be revealed by differing lighting and vegetation conditions and to delineate changes in the man-made and natural conditions, including slide development.

Barredo et al. (2000) employed the preceding GIS-assisted direct and indirect heuristic multicriteria evaluation procedure to evaluate hazards from the Barranco de Tirajana basin, composed of several large landslides. They revealed that the above techniques are relatively simple and cost-effective for the landslide hazard at medium scales, in particular, when costly geotechnical and groundwater data are nor readily available.

Landslide hazard has been defined (Varnes and International Union of Geological Sciences , 1984) as "the probability of occurrence within a specified period of time and within a given area of a potentially damaging phenomenon." Furthermore, Varnes and International Union of Geological Sciences (IUGS, 1984) described the landslide risk assessment as "the expected degree of loss due to a landslide (specific risk) and the expected number of lives lost, people injured, damage to property and disruption of economic activity (total risk)." As shown in Figure 4, the integrated assessment of landslide hazard and risk requires a broad-based knowledge from a wide spectrum of disciplines including geosciences, geomorphology, meteorology, hydrogeology, and geotechnical engineering (Chowdhury et al., 2001).

Landslide hazards are commonly delineated on inventory maps, which display distributions of hazard classes and identify areas that potential landslides may be generated (Leroi, 1996). Inventory maps show the location and, where applicable, the date of occurrence and historical records of landslides in a region (Hansen, 1984 and Wieczorek, 1984). These maps are prepared by different techniques, and, ideally, provide information concerning the spatial and temporal probability, type, magnitude, velocity, runout distance, and retrogression limit of the mass movements predicted in a designated area (Hartlen and Viberg, 1988). Details of inventory maps depend on available resources and are based on the scope, extent of study area, scales of base maps, aerial photographs, and future land use (Guzzetti, F., 2003).

Evidently, the extent of information required concerning the landslide hazard analysis will depend on the level and nature of proposed development for a region. The negligence of incorporating the impact of potential landslide activity on a project or the prospects of new development on landslide potential may lead to increased risk (Primer, 1991). The assessment of future landslide susceptibility in a region will require the evaluation of relevant conditions and processes controlling landslides, these being causative factors. These conditions and processes include historical events, slope geometry, bedrock characteristics, and hydrologic features of the designated area. Brabb (1984) indicated that the results of risk assessments for large areas can be expressed in the form of landslide hazard or landslide risk maps. Einstein (1997 and 1988) has presented a comprehensive mapping procedure for landslide management. Following are key features of mapping procedures proposed by Einstein (1997 and 1988):

♦ *State-of-nature maps* – are used to characterize a site, present data without interpretation, such as geologic and topographic maps, precipitation data, and results of site investigation.

♦ *Danger maps* – are utilized to identify the failure modes involving debris flows, rock falls, etc.

♦ *Hazard maps* – are employed to exhibit the probability of failure related to the possible modes of failure on danger maps. Alternatively, the results can be expressed qualitatively as high, medium, or low.

♦ *Management maps* – are incorporated to entail summaries of management decisions.

Furthermore, following scales of analyses for landslide hazard zonations have been outlined by the International Association of Engineering Geology (Soeters and van Westen, 1996):

♦ National Scale (<1:1,000,000) – This is a low-level detail map intended to provide a general inventory of nationwide hazard. It is used to notify national policy makers and general public.

♦ Regional Scale (1:100,000 to 1:500,000) – Because landslide hazards are considered to be undesirable factors as far as the planners are concerned, the regional mapping scale is employed in evaluating possible constraints due to instability related to the development of large engineering projects and regional development plans. In general, these types of maps are constructed in early phases of regional development projects with low–level details and cover large study areas, on the order of 1000 km^2 or more. They are used to identify areas where landsliding could be a constraint concerning the development of rural or urban transportation projects. "Terrain units with an areal extent of several tens of hectares are outlined and classified according to their susceptibility to occurrence of mass movements," as stipulated by Soeters and van Westen (1996).

♦ Medium Scale (1:25,000 to 1:50,000) – This range is considered to be a suitable scale range for landslide hazard maps (Primer, 1991). As such, they are utilized to identify the hazard zones in developed areas with large structures, roads, and urbanization. Considerably greater levels of detail are required to prepare the maps at this scale and the details should encompass slopes in adjacent sites in the same lithology with the possibility of having different hazard scores depending on their characteristics. Furthermore, distinction should be made between various slope segments, located within the same terrain unit, such as rating of a concave slope as opposed to a convex slope (Soerters and van Westen, 1996).

♦ Large Scale (1:5,000 to 1:15,000) – Maps of this scale are generally prepared for limited areas based on both interpretation of aerial photographs and extensive field investigations that utilize various techniques applied in routine geotechnical engineering, engineering geology and geomorphology (Guzzetti et al., 2000).

Guzzetti (2003) has made the following recommendations concerning the preparation and use of landslide inventory maps. He stated that the landslide cartography should be increasingly utilized and landslide inventory maps should be created for entire regions based on consistent and reproducible methods. On a regional scale, Carrara et al. (1991) indicated: "the temporal dimension of landsliding is essentially a function of the triggering mechanisms which are climatic (due to extreme rainfall) or geodynamic (earthquakes) in nature." Thus, it would be advantageous to prepare landslide inventory maps following each landslide-triggering event such as a rainstorm, a snowmelt event, or an earthquake, for the entire affected region. These inventory maps will provide valuable information regarding types, extent and severity of damage caused by the event and will help in assessing the impact of the landslide events on infrastructures in that region.

It is important that the quality, reliability, and sensitivity of landslide hazard models and maps be carefully examined. Guzzetti (2003) has suggested that the created models need to be checked against high-quality inventory maps and reliable historical catalogues of landslide events. It has also been recommended (Guzzetti, 2003) that both quantitative and qualitative methods be increasingly performed in regard to total risk assessments at local and regional scales. There is a lack of sufficient data and case histories for critical evaluation of techniques and models concerning regional and local landslide risk assessments. Thus, there is a need for compilation of relevant data that will help to compare qualitative and quantitative risk assessment procedures and outcomes. It is anticipated that the recent development of statistical analyses utilizing GIS techniques will enhance analyses of spatial data sets and, thus, quantitative representation of landslide potential along with graphical depictions (Carrara and Guzzetti, 1995).

5. Landslide Risk Assessment

By and large, the elements at risk involve property, people, services, such as water supply or drainage or electricity supply, roads and communication facilities, and vehicles on roads. The consequences may not, however, be limited to property damage and injury/loss of life. Other factors include public outrage, political effects, loss of business confidence, effect on reputation, social upheaval, and consequential costs, such as litigation (Australian Geomechanics Society, 2000). Many of these may not be readily quantifiable and will require considerable judgment if they are to be included in the assessment. Consideration of such consequences may constitute part of the risk evaluation process by the client/owner/regulator.

Risk estimation may be carried out quantitatively, semi quantitatively, or qualitatively. The quantitative approach is explained below to illustrate the principles involved. Wherever possible, the risk estimate should be based on a quantitative analysis, even though the results may be summarized using qualitative terminology. Quantitative risk estimation involves integration of the frequency analysis and the consequences.

For property, the risk can be calculated from (Australian Geomechanics Society, 2000):

$$R(Prop) = P(H) \times P(S:H) \times V(Prop:S) \times E \tag{1}$$

where:

R(Prop) is the risk (annual loss of property value).
P(H) is the annual probability of the hazardous event (the landslide).
P(S:H) is the probability of spatial impact by the hazard (i.e., probability of the landslide impacting the property, taking into account the travel distance) and for vehicles, for example, the temporal probability.

V(Prop:S) is the vulnerability of the property to the spatial impact (proportion of property value lost).

E is the element at risk (e.g. the value or net present value of the property).

For loss of life, the individual risk can be calculated from (Australian Geomechanics Society, 2000):

$$R(DI) = P(H) \times P(S:H) \times P(T:S) \times V(D:T) \tag{2}$$

where:

R(DI) is the risk (annual probability of loss of life (death) of an individual).

P(H) is the annual probability of the hazardous event (the landslide).

P(S:H) is the probability of spatial impact by the hazard (e.g., of the landslide impacting a building (location) taking into account the travel distance) given the event.

P(T:S) is the temporal probability (e.g., of the building being occupied by the individual) given the spatial impact.

V(D:T) is the vulnerability of the individual (probability of loss of life of the individual given the impact).

A complete risk analysis involves consideration of all landslide hazards at a designated site (e.g., large, deep seated landsliding, smaller slides, rock fall, debris flows) and relevant elements at risk. For total risk, in relation to the property and/or life, the risk for each hazard of each element is summed. Because estimates made for an analysis will be imprecise, sensitivity analyses are useful in evaluating the effects of changing assumptions or estimates. Variation in the estimate of risk by one or two orders of magnitude, or perhaps three orders of magnitude at low risks, will not be uncommon. The resulting sensitivity may aid judgment as to the critical aspects requiring further investigation or evaluation.

6. Landslide Risk Treatment

6.1 Landslide Risk Treatment Options

Risk treatment is the final stage of the risk–management process and provides the methodology for controlling the risk. At the end of the evaluation procedure, it is up to the client or to policy makers to decide whether to accept the risk or not, or to decide that more detailed study is required. The landslide risk analyst can provide background data or normally acceptable limits as guidance to the decision maker but should not be making the decision. Part of the specialist's advice may be to identify the options and methods for treating the risk. Typical options would include (Australian Geomechanics Society, 2000):

- *Accept the risk* - This will usually require the risk to be considered to be within the acceptable or tolerable range.

- *Avoid the risk* - This will entail avoiding the project, thus seeking an alternative site or form of development so that the revised risk becomes acceptable or tolerable.

- *Reduce the likelihood* - This requires stabilization measures to control the initiating circumstances, such as re-profiling the surface geometry, or installing groundwater drainage, anchors, stabilizing structures, protective structures, etc.

- *Reduce the consequences* - This requires provision of defensive stabilization measures, amelioration of the behavior of the hazard, or relocation of the development to a more favorable location to achieve an acceptable or tolerable risk.

- *Monitoring and warning systems* - In some situations, monitoring (such as by regular site visits, or by surveys), and the establishment of warning systems may be used to manage the risk on an interim or permanent basis. Monitoring and warning systems may be regarded as another means of reducing the consequences.

- *Transfer the risk* – This requires that either another authority to accept the risk or to compensate for the risk such as by insurance.

- *Postpone the decision* - If there is sufficient uncertainty, it may not be appropriate to make a decision on the data available. Further investigation or monitoring will be required to provide data for better evaluation of the risk.

The relative costs and benefits of various options need to be considered so that the most cost effective solutions, consistent with the overall needs of the client, owner and regulator, can be identified. Combinations of options or alternatives may be appropriate, particularly where relatively large reductions in risk can be achieved for relatively small expenditures. Prioritization of alternative options is likely to assist with selection.

6.2 Landslide Remedial Measures

Correction of an existing landslide or the prevention of a pending landslide is a function of reduction of the driving forces or increase in the available resisting forces. Any remedial measure used must involve one or both of the above parameters. Many general reviews of the methods of landslide remediation have been made. The interested reader is particularly directed to Hutchinson (1977), Zaruba and Mencl (1982), Bromhead (1992) and Fell (1994), Schuster (1995). IUGS WG/L (Popescu, 2001) has prepared a short checklist of landslide remedial measures arranged in four practical groups, namely: modification of slope geometry, drainage, retaining structures and internal slope reinforcement. A flow diagram (Figure 5) exhibits the sequence of various phases involved in the planning, design, construction and monitoring of remedial works (Kelly and Martin, 1986).

Hutchinson (1977) has indicated that drainage is the principal measure used in the mitigation of landslides, with modification of slope geometry the second most used method. These are also generally the least costly of the four major categories, which is obviously why they are the most used. The experience shows that while one remedial measure may be dominant, most landslide repairs involve use of a combination of two or more of the major categories. For example, while restraint may be the principal measure used to correct a particular landslide, drainage and modification of slope geometry, to some degree and by necessity, are also utilized.

Drainage is often a crucial remedial measure due to the important role played by pore-water pressure in reducing shear strength. Because of its high stabilization efficiency in relation to cost, drainage of surface water and groundwater is the most widely used, and generally the most successful stabilization method. As a long-term solution, however, it suffers greatly because the drains must be maintained if they are to continue to function (Bromhead, 1992).

Surface water is diverted from unstable slopes by ditches and pipes. Drainage of shallow groundwater is usually achieved by networks of trench drains. Drainage of the failure surfaces, on the other hand, is achieved by counterfort or deep drains which are trenches sunk into the ground to intersect the shear surface and extending below it. In the case of deep landslides, often the most effective way of lowering groundwater is to drive drainage adits into the intact material beneath the landslide. From this position, a series of upward–directed drainage holes can be drilled through the roof of the tunnel to drain the sole of the landslide. Alternatively, the adits can connect a series of vertical wells sunk down from the ground surface. In instances where the groundwater is too deep to be reached by ordinary trench drains and where the landslide is too small to justify an expensive drainage adit or gallery, bored sub-horizontal drains can be used. Another approach is to use a combination of vertical drainage wells linked to a system of sub-horizontal borehole drains.

Modification of slope geometry is a most efficient method, particularly for deep–seated landslides. However, the success of corrective slope regrading (fill or cut) is determined not merely by size or shape of the alteration, but also by position on the slope. Hutchinson (1977) provided details of the "neutral line" method to assist in finding the best location to place a stabilizing fill or cut. There are some situations where this approach is not simple to adopt. These include long translational landslides where there is no apparent toe or crest and situations where the geometry of the unstable mass is complex and a change in topography that improves the stability of one area may reduce the stability of another.

Schuster (1995) discussed recent advances in the commonly used drainage systems, while briefly mentioning less commonly used, but innovative, means of drainage, such as electro-osmotic dewatering, and vacuum and siphon drains. In addition, buttress counterforts of coarse-grained materials placed at the toes of unstable slopes often are successful as remedial measures.

During the early part of the post-world war II period, landslides were generally seen to be "engineering problems" requiring "engineering solutions" involving correction by the use of structural techniques. This structural approach initially focused on retaining walls, but has subsequently been diversified to include a wide range of more sophisticated techniques including passive piles and piers, cast-in-situ reinforced concrete walls, and reinforced–earth retaining structures. When properly designed and constructed, these structural solutions can be extremely valuable, especially in areas with high loss potential or in restricted sites. However fixation with structural solutions has in some cases resulted in the adoption of overly expensive measures that have proven to be less appropriate than alternative approaches involving slope geometry modification or drainage (Department of the Environment, 1994).

Over the last several decades, there has been a notable shift toward "soft engineering," non-structural solutions, including classical methods such as drainage and modification of slope geometry, but also some novel methods such as lime/cement stabilization, grouting or soil nailing (Powell, 1992). The cost of non-structural remedial measures is considerably lower than the cost of structural solutions. In addition, structural solutions, such as retaining walls, involve exposing the slope during construction and often require steep temporary excavations. Both of these operations increase the risk of failure during construction for over-steepening or increased infiltration from rainfall. In contrast, the use of soil nailing as a non-structural solution to strengthen the slope avoids the need to open or alter the slope from its current condition.

Environmental considerations have increasingly become an important factor in the choice of suitable remedial measures, particularly issues such as visual intrusion in scenic areas or the impact on nature or geological conservation interests. An example of a "soft engineering" solution, more compatible with the environment, is the stabilization of slopes by the combined use of vegetation and man-made structural elements working together in an integrated manner known as biotechnical slope stabilization (Schuster, 1995). The basic concepts of vegetative stabilization are not new - vegetation has a beneficial effect on slope stability by the processes of interception of rainfall, and transpiration of groundwater, thus maintaining drier soils and enabling some reduction in potential peak groundwater pressures. In addition to these hydrological effects, vegetation roots reinforce the soil, increasing soil shear strength, while tree roots may anchor into firm strata, providing support to the upslope soil mantle and buttressing and arching. A small increase in soil cohesion induced by the roots has a major effect on shallow landslides. The mechanical effect of vegetation is not significant for deeper seated landslides, while the hydrological effect is beneficial for both shallow and deep landslides. However, vegetation may not always assist slope stability. Destabilizing forces may be generated by the weight of the vegetation acting as a surcharge and by wind forces acting on the exposed vegetation, although both of these are very minor effects. Roots of vegetation may also act adversely by penetrating and dilating the joints of widely jointed rocks. For detailed information on research into the engineering role of vegetation for slope stabilization refer to Greenway (1987) and Wu (1991). In addition, the "Geotechnical Manual for Slopes" (Geotechnical Control Office, 1981) includes an excellent table noting the hydrological and mechanical effects of vegetation.

The concept of biotechnical slope stabilization is generally cost effective as compared to the use of structural elements alone; it increases environmental compatibility, and allows the use of local natural materials. Interstices of the retaining structure are planted with vegetation whose roots bind together the soil within and behind the structure. The stability of all types of retaining structures with open gridwork or tiered facings benefits from such vegetation. An example of a composite vegetated geotextile/geogrid reinforced structure named "Biobund" was presented by Barker (1991).

6.3 Levels of Effectiveness and Acceptability That May Be Applied in the Use of Remedial Measures

Terzaghi (1950) stated that, "if a slope has started to move, the means for stopping movement must be adapted to the processes which started the slide". For example, if erosion is a causal process of the slide, an efficient remediation technique would involve armoring the slope against erosion, or removing the source of erosion. An erosive spring can be made non-erosive by either blanketing with filter materials or drying up the spring with horizontal drains, etc.

The greatest benefit in understanding landslide-producing processes and mechanisms lies in the use of the above understanding to anticipate and devise measures to minimize and prevent major landslides. The term major should be underscored here because it is neither possible nor feasible, nor even desirable, to prevent all landslides. There are many examples of landslides that can be handled more effectively and at less cost after they occur. Landslide avoidance through selective locationing is obviously desired - even required - in many cases, but the dwindling number of safe and desirable construction sites may force more and more the use of landslide–susceptible terrain.

Selection of an appropriate remedial measure depends on: a) engineering feasibility, b) economic feasibility, c) legal/regulatory conformity, d) social acceptability, and e) environmental acceptability. A brief description of each aspect is presented herein:

a) Engineering feasibility involves analysis of geologic and hydrologic conditions at the site to ensure the physical effectiveness of the remedial measure. An often-overlooked aspect is being certain that the design will not merely divert the problem elsewhere.

b) Economic feasibility takes into account the cost of the remedial action as composed to the benefits it provides. These benefits include deferred maintenance, avoidance of damage (including loss of life), and other tangible and intangible benefits.

c) Legal-regulatory conformity provides for the remedial measure meeting local building codes, avoiding liability to other property owners, and related factors.

d) Social acceptability is the degree to which the remedial measure is acceptable to the community and neighbors. Some measures for a property owner may prevent further damage but be an unattractive eyesore to neighbors.

e) Environmental acceptability addresses the need for the remedial measure to not adversely affect the environment. De-watering a slope to the extent it no longer supports a unique plant community may not be environmentally acceptable solution.

Just as there are a number of available remedial measures, so are there a number of levels of effectiveness and levels of acceptability that may be applied in the use of these measures. We may have a landslide, for example, that we choose to live with. Although this type of landslide poses no significant hazard to the public, it will require periodic maintenance through removal, due occasional encroachment onto the shoulder of a roadway. The permanent closure of the Manchester–Sheffield road at Mam Tor in 1979 (Skempton et al., 1989) and the decision not to reopen the railway link to Killin following the Glen Ogle rockslide in U.K. (Smith, 1984) are well–known examples of abandonment due to the effects of landslides in which repair was considered uneconomical.

Most landslides, however, usually must be dealt with sooner or later. How they are handled depends on the processes that prepared and precipitated the movement, the landslide type, the kinds of materials involved, the size and location of the landslide, the place or components affected by or the situation created as a result of the landslide, available resources, etc. The technical solution must be in harmony with the natural system, otherwise the remedial work will be either short–lived or excessively expensive. In fact, landslides are so varied in type and size, and in most instances, so dependent upon special local circumstances, that for a given landslide problem there is more than one method of prevention or correction that can be successfully applied. The success of each measure depends, to a large extent, on the degree to which the specific soil and groundwater conditions are prudently recognized in an investigation and incorporated in design.

The failure to properly recognize and apply the above conditions could result in the failure of remedial works, as illustrated by the main landslide event at Craco, South Italy (Del Prete and Petley, 1982). A retaining wall built on lines of contiguous piles installed to arrest the movement of a reactivated ancient landslide at Craco actually caused a dramatic increase in landslide activity and the final collapse due to a combination of the following three unfavorable effects: (i) the retaining wall was founded on piles which did not penetrate beyond the surface of sliding; (ii) the wall was located in a position where additional loading caused a reduction in the stability condition; (iii) the impervious wall acted as a dam to the flow of groundwater, resulting in an increase in pore–water pressure behind the contiguous piles.

As many of the geological features, such as sheared discontinuities are not known in advance, it is more advantageous to plan and install remedial measures on a "design–as–you–go basis." That is, the design has to be flexible enough to accommodate changes during or subsequent to the construction of remedial works.

7. Landslide Monitoring

Monitoring of landslides plays an increasingly important role in the context of living and coping with these natural hazards. The classical methods of land surveys, inclinometers, extensometers and piezometers are still the most appropriate monitoring measures. In the future, the emerging techniques based on remote sensing and remote access techniques will undoubtedly be of main interest (Laboratoire Central des Ponts et Chaussées, 1994).

Department of Environment (1994) has identified the following categories of monitoring, designed for slightly differing purposes but generally involving similar techniques:

1. Preliminary monitoring involves provision of data on pre-existing landslides so that the dangers can be assessed and remedial measures can be properly designed or the site be abandoned.

2. Precautionary monitoring is carried out during construction in order to ensure safety and to facilitate redesign, if necessary.

3. Post-construction monitoring is considered in order to check on the performance of stabilization measures and to focus attention on problems that require remedial measures.

Observational methods based on careful monitoring – before, during and after construction – are essential in achieving reliable and cost–effective remedial measures (Brandl, 1995).

8. Landslide Warning Systems

When dealing with a slope of precarious stability and/or presenting a risk which is considered too high, a possible option is to do nothing in regard to mitigation, but to install a warning system in order to insure or improve the safety of people. It is worth noting that warning systems do not modify the hazard but contribute to reducing the consequences of the landslide and thus the risk, in particular the risk associated to the loss of life.

Leroueil (1996) defined the following four possible different stages of landslide activity:

1. Pre-failure stage, when the soil mass is still continuous. This stage is mostly controlled by progressive failure and creep;

2. Onset of failure characterized by the formation of a continuous shear surface through the entire soil or rock mass;

3. Post-failure stage, which includes movement of the soil or rock mass involved in the landslide from just after failure until it essentially stops;

4. Reactivation stage when the soil or rock mass slides along one or several pre-existing shear surfaces. This reactivation can be occasional or continuous with seasonal variations of the rate of movement.

Various types of warning systems have been proposed and the selection of an appropriate one should take into account the stage of landslide activity:

1. At pre-failure stage, the warning system can be applied either to revealing factors or to triggering or aggravating factors. Revealing factors can be, for example, the opening of fissures or the movement of given points on the slope; in such cases, the warning criterion will be the magnitude or rate of movement. When the warning system is associated with triggering or aggravating factors, there is a need to first define the relation between the magnitude of factors controlling the stability condition or the rate of movement of the slope. The warning criterion can be a given hourly rainfall or the cumulative rainfall during a certain period of time, increased pore–water pressure, a given stage of erosion, a minimum negative pore pressure in a loess deposit, etc.

2. At failure stage, the warning system can only be linked to revealing factors, generally a sudden acceleration of movements or the disappearance of a target.

3. At post-failure stage, the warning system has to be associated to the expected consequences of the movement. It is generally associated with the rate of movement and run out distance.

The majority of remedial measures, outlined above, can be cost-prohibitive and may be socially and politically unpopular. As a result, there may be a temptation to adopt and rely instead upon the installation of apparently cheaper and much less disruptive monitoring and warning systems to "save" the population from future catastrophes. However, for such an approach to be successful it is necessary to fulfill satisfactorily each of the following steps (Hutchinson, 2001):

a) The monitoring system shall be designed to record the relevant parameters, to be in the right places, and to be sound in principle and effective in operation.

b) The monitoring results need to be assessed continuously by suitable experts.

c) A viable decision shall be made, with a minimum of delay, that the danger point has been reached.

d) The decision should be passed promptly to the relevant authorities, with a sufficient degree of confidence and accuracy regarding the forecast place and time of failure for those authorities to be able to act without fear of raising a false alarm.

e) Once the authorities decide to accept the technical advice, they must pass the warning on to the public in a way that will not cause panic and possibly exacerbate the situation,

f) The public needs to be well–informed and prepared in advance in order to respond in an orderly and pre-arranged manner.

In view of the preceding discussion, it is not surprising that, although there have been a few successes with monitoring and warning systems, particularly in relatively simple, site-specific situations, there have been many cases where these have failed, because one or more of requirements a) through f) above have been violated, often with tragic and extensive loss of life. It is concluded, therefore, that sustained good management of an area, as outlined above, should be our primary response to the threat of landslide hazards and risks, with monitoring and warning systems being in a secondary, supporting role.

9. Forecasting the Time of Landslides

Landslides are very complex phenomena and are difficult to predict. They involve materials ranging over many orders of magnitudes in size, from fine–grained particles to masses of earth/rock of several cubic kilometers (Hamilton, 1997). The velocity of mass movements also varies over a wide range, from creeping movements of mm/yr to extremely rapid avalanches that travel at several hundred km/hr (Cruden and Varnes, 1996). Moreover, they span the geologic-hydrologic interface from completely dry materials to viscous fluid–type flows (Hamilton, 1997). As a result, forecasting the time of landslides remains a crucial and still an unresolved problem (Guzzetti, 2003).

Landslide prediction can be classified as long–term, intermediate–term, or short–term (Hamilton, 1997). Long–term prediction of landslides is typically attained via landslide hazard maps, which are actually susceptibility maps, for large areas. As mentioned previously, these maps contribute to assessments of long-term characteristics and warning of landslide hazards; hence, they provide a framework for identifying the need for additional data, and effective mitigation techniques, along with zoning or land use planning (United Nations, 1996). Evidently, as more climatological and seismological events concerning the recurrence of storms and earthquakes are incorporated, the time-related element will be integrated allowing determination of the time-dependent changes in the probability of landsliding (Campbell et al. 1998). It is anticipated that on-going research studies regarding factors affecting distribution and mechanisms of storm-induced or earthquake-triggered landslides will result in future refinements of criteria being used to prepare landslide hazard maps along with integration of time elements (Sassa, 1996).

Landslide monitoring is considered to provide the necessary data that can be used for intermediate-term prediction. Appearance of cracks, fluctuation of moisture in soils, and acceleration of surface or subsurface movements provide precursory evidence of landslide movements (Saito, 1965; Fukuzono, 1985). Specifically, the acceleration of surface or subsurface movements enables the most direct detection of impending landsliding (Voight and Kennedy, 1979).

Monitoring, described above, entails compilation of meteorological, hydrological, topographical, and geophysical data (Mikkelsen, 1996). The advent of automatic sampling, recording, and transmitting devices has enabled practical prediction of landslide movements (Hamilton, 1997). Although prediction of landslide movement, based on interaction between climate and slope movement, is a daunting task at this time, it may become more viable in the future due to on-going research and monitoring of regional weather patterns. Wilson et al. (1995) has determined that the critical amount of rainfall required to trigger flows or other shallow landslides depend upon the climate. Consequently, the intermediate-term prediction of landslide movements based on rainfall thresholds will necessitate compilation of detailed regional meteorological data.

Among approaches to the mitigation of landslide risk, the prediction of the time of occurrence for a first time landslide deserves special consideration. The task is far from being simple because the fundamental physics controlling the nature and shape of the creep curve of geomaterials has not been fully elucidated yet. Moreover, all the relevant parameters and boundary conditions are not clearly defined and it is impossible to forecast the triggering factors originating outside the sliding mass (for example, heavy rainfall). An important key to the prediction of landslide failure time should be the stress-strain-time relations, but the heterogeneity of the geological conditions, groundwater seepage conditions, associated pore–water pressures on the potential sliding surface and scale effects make the laboratory evaluation of the geomechanical parameters barely adequate for the simulation of the temporal evolution of a potential slide using numerical models (Federico et al., 2002; Fell et al., 2000).

Several methods have been proposed for the prediction concerning the time of occurrence of landslides. In engineering practice, such methods, that infer the time to failure by means of monitored surface displacements, are preferred for a prediction, given that they remove all uncertainties involved in these problems. One of the first, most spectacular and well documented predictions of slope failure, based upon displacement monitoring, was carried out at the Chuquicamata mine in Chile (Kennedy and Niermeyer, 1970): The date of failure was exactly predicted by means of a rough extrapolation of displacement data. Hoek and Bray (1977) pointed out, the circumstance is not of great importance; in fact, from the point of view of an engineer, even a prediction with an error of few weeks is reasonable and helps in making decisions. As a consequence, one may state that the key to the prediction is the correct choice and a good monitoring of the relevant physical factors, rather than the principle selected for inferring the time to failure.

Regardless of the technique used for extrapolating the time to failure, the quality of the prediction depends on the quality of the data, so that a clear identification of the critical points or variables selected for monitoring is strongly required in order to get a consistent prediction. This entails the need for developing of an understanding of pre-failure deformations and other precursory signs of different landslides mechanisms (Fell et al., 1997). Accordingly, the help offered by slope monitoring methods, particularly GPS and Time Domain Reflectometry (TDR), can be noticeable. For some methods, the frequency of observation seems to condition the effectiveness of the prediction, as well as the extent of the time span of data collection (i.e., the monitoring system should be installed as soon as possible). The observation needs also to be extended to other parameters, different than displacements, such as pore pressure or crack aperture. An expert's visual examination of plots of the monitored data allows assessment of the present creep stage of the slope. If this stage is one of tertiary creep, the slope is in a critical condition, close to the failure, and the confidence of the prediction is required to be greater.

10. Concluding Remarks

Assessing the landslide hazard is the most important step in landslide risk management. Once that has been done, it is feasible to assess the number, size, and vulnerability of the fixed elements at risk (structures, roads, railways, pipelines, etc.), and thence the damage they will suffer. The various risks have to be combined to arrive at a total risk in financial terms. Comparison of this with, for instance, cost-benefit studies of the cost of relocation of populations and facilities, or mitigation of the hazard by countermeasures, provides a useful tool for management and decision-making.

In many parts of the world there exist numerous sites where there is undue risk from landslides to communities and infrastructure. The urgent need is to identify and rank these sites, using well–established methods of landslide hazard and landslide risk analysis, and then to mitigate these risks appropriately and effectively. The necessary actions should be taken as soon as possible, while there is yet time.

It should be emphasized that these include not only various direct measures, such as relocation of dwellings and infrastructure or slide stabilization but also "good housekeeping" of the region as a whole, as for example sustained, ecologically sensitive management of land use, sound planning, obtaining information, making emergency arrangements, etc.. In Hong Kong such approaches have had dramatic success, reducing the average rate of landslide fatalities per year per person to 5×10^{-7}, a tenth of what it was before the introduction of a slope–safety regime (through what is now the Geotechnical Engineering Office) in late 1972 (Malone, 1998).

A pragmatic approach of living with landslides and reducing the impact of landslide problems in urban areas is well illustrated by the strategy adopted to cope with landslide problems at Ventnor, Isle of Wight, U. K. (Lee et al., 1991). Ventnor is an unusual situation in that the entire town lies within an ancient landslide complex. The spatial extent and scale of the problems at Ventnor has indicated that total avoidance or abandonment of the site are out of question and large-scale conspicuous engineering structures would be unacceptable in a town dependent on tourism. Instead coordinated measures have been adopted to limit the impacts of human activity that promote ground instability by planning control, control of construction activity, preventing water leakage and improving building standards. In addition, good maintenance practice by individual homeowners proved to be a significant help, because neglect could have resulted in localized instability problems.

Much progress has been made in developing techniques to minimize the impact of landslides, although new, more efficient, quicker and cheaper methods could well emerge in the future. There are a number of levels of effectiveness and levels of acceptability that may be applied in the use of these measures, for, while one slide may require an immediate and absolute long-term correction, another may only require minimal control for a short period.

Whatever the measure chosen, and whatever the level of effectiveness required, the geotechnical engineer and engineering geologist have to combine their talents and energies to solve the problem. Solving landslide–related problems is changing from what has been predominantly an art to what may be termed an art-science. The continual collaboration and sharing of experience by engineers and geologists   will no doubt move the field as a whole closer toward the science end of the art–science spectrum than it is at present.

REFERENCES

Australian Geomechanics Society (AGS), Sub-Committee on Landslide Risk Management (2000). *Landslide Risk Management Concepts and Guidelines*, pp-49-92

Barker, D.H. (1991). "Developments in Biotechnical Stabilization in Britain and the Commonwealth," *In Proc., Workshop on Biotechnical Stabilization*, University of Michigan, Ann Arbor, A-83 to A-123.

Barredo, J.I., Benavides, A., Hervás, J. and Van Westen, C.J., (2000). "Comparing Heuristic Landslide Hazard Assessment Techniques Using GIS in the Tirajana Basin, Gran Canaria Island, Spain," *International Journal of Applied Earth Observation and Geoinformation*, v. 2 (1), 9-23.

Brabb, E.E. (1984), "Innovative approaches to landslide hazard and risk mapping," *In Proceedings of the Fourth International Symposium on Landslides*, Canadian Geotechnical Society, Toronto, Canada, v. 1: 307-323.

Brandl, H. (1995). "Observational Method in Slope Engineering," *In Proc., Intern. Symp. 70 Years of Soil Mechanics*, Istanbul, 1-12.

Bromhead, E.N. (1992). "The Stability of Slopes," *Blackie Academic & Professional*, London.

Campbell, R.H., Bernknopf, R.L., and Soller, D.R. (1998), "Mapping Time-Dependent Changes in Soil Slip-Debris Flow Probability," *U.S. Geological Survey Miscellaneous Investigations Series Map* I-2586.

Carrara A., Cardinali M., Detti R., Guzzetti G., Pasqui V. and Reichenbach P. (1991). "GIS Techniques and Statistical Models in Evaluating Landslide Hazard. Earth Surface Processes and Landforms," v. 16, p. 427-445.

Carrara, A., and F. Guzzetti (eds). (1995) "Geographical Information Systems in Assessing Natural Hazards," Kluwer Academic Publishers, Dordrecht, Netherlands, 353 p.

Chowdhury, R., Flentje, P., and Ko Ko, C. (2001) "A Focus on Hilly Areas Subject to the Occurrence and Effects of Landslides," *Global Blueprint for Change*, 1st Edition – Prepared in conjunction with the International Workshop on Disaster Reduction, August 19-22, 2001.

Crozier, M.J. (1986). "Landslides - Causes, Consequences and Environment," *Croom Helm*, London.

Cruden,D.M. (1991). "A Simple Definition of a Landslide," *Bulletin International Association of Engineering Geology (IAEG)*, 43:27-29.

Cruden, D.M. (1997). "Estimating the Risks from Landslides Using Historical Data," *in Landslide Risk Assessment*, Cruden, D.M., and Fell, R. (eds.), Balkema, pp. 277-284.

Cruden, D.M. and Varnes, D.J. (1996). "Landslide types and processes: in "Landslides Investigation and Mitigation," K. Turner, and R.L. Schuster (eds.) *Transportation Research Board Special Report 247*, National Research Council, Washington, DC.

Del Prete, M., and Petley, D.J. (1982). "Case History of the Main Landslide At Craco, Basilicata, South Italy," *Geologia Applicata e Idrogeologia, XVII*: 291-304.

Department of the Environment (1994). D.K.C. Jones and E.M. Lee (eds.) "Landsliding in Great Britain", *Her Majesty's Stationary Office*, London.

Einstein, H.H. (1988). "Landslide Risk Assessment Procedure," *In Proceedings of 5th International Symposium on Landslides*, Lausanne, Switzerland, A.A., Balkema, Rotterdam, Netherlands, v. 2, pp.1075-1090.

Einstein, H.H. (1997). "Landslide Risk – Systematic Approaches to Assessment and Management," *in Landslide Risk Assessment*, Cruden, D.M., and Fell, R. (eds.), Balkema, Rotterdam, pp. 25-50.

Einstein, H.H., Karam, K.S. (2001). "Risk Assessment and Uncertainties," *In Proceedings of the International Conference on Landslides* – Causes, Impacts and Countermeasures," Davos, Switzerland, pp. 457-488.

Federico, A., Fidelibus, C. Interno, G. (2002). "The Prediction of Landslide Time to Failure – A State-of-the-Art," *In Proc. Int. Conf. Landslides, Slope Stability and Safety of Infrastructures*, Singapore, pp. 167-180.

Fell, R. (1994). "Stabilization of Soil and Rock Slopes," *In Proc. East Asia Symp. and Field Workshop on Landslides and Debris Flows*, Seoul, 1:7-74.

Fell, R. (1994). "Landslide Risk Assessment and Acceptable Risk," *Canadian Geotechnical Journal*, 31, pp.261-272.

Fell, R. and Hartford, D. (1997). "Landslide Risk Management," *In Landslide Risk Assessment*, Cruden, D.M., and Fell, R. (eds.), Balkema, Rotterdam, pp. 51-110.

Fell, R., Hungr, O., Leroueil, S. and Riemer, W. (2000). Keynote Lecture – "Geotechnical Engineering of the Stability of Natural Slopes, and Cuts and Fills in Soil," *In Proc. GeoEng 2000,* Melbourne, pp. 21-120.

Fukuzono, T. (1985). "A New Method for Predicting the Failure Time of a Slope." *In Proceedings of the 4th International Conference and Field Workshop in Landslides*, Tokyo, pp.145-150.

Geotechnical Control Office (1981). "Geotechnical Manual For Slopes," *Public Works Department*, Hong Kong.

Greenway, D.R. (1987). "Vegetation And Slope Stability," *Slope Stability*, John Wiley & Sons Ltd., NY, pp. 187 - 230.

Guzzetti, F. (2003). "Landslide Cartography, Hazard Assessment and Risk Evaluation: Overview, Limits and Prospective," http://www.mitch-ec.net/workshop3/Papers/paper_guzzetti.pdf

Guzzetti, F., Cardinali, M., Reichenbach, P., and Carrara, A. (2000). "Comparing Landslide Maps: A Case Study in the Upper Tiber River Basin, Central Italy." *Env. Management*, 25:3, pp. 247-363.

Hamilton, R. (1997). "Report on Early Warning Capabilities for Geological Hazards," *International Decade for Natural Disaster Reduction (IDNDR) Secretariat*, Geneva, http://www.unisdr.org/unisdr/docs/early/geo/geo.htm, Accessed, October, 8, 2003.

Hansen, A. (1984). "Landslide Hazard Analysis", In Brunsden, D. and Prior, D.B., (eds.), *Slope Instability*, John Wiley & Sons, New York, pp. 523-602.

Hartlen, J. and Viberg, L. (1988). "General Report: Evaluation of Landslide Hazard," *In Proceedings of 5th International Symposium on Landslides*, Lausanne, A.A. Balkema, Rotterdam, Netherlands, v. 2, pp. 1037-1057.

Hutchinson, J.N. (1977). "The Assessment of The Effectiveness of Corrective Measures in Relation to Geological Conditions and types Of Slope Movement." *Bulletin of International Association of Engineering Geology (IAEG)*, 16: 131-155.

Hutchinson J. N. (2001). "Landslide Risk - to know, to Foresee, to Prevent," *Journal of Technical and Environmental Geology*, n. 3, pp. 3-22.

Hoek E. and Bray J. (1977). "Rock Slope Engineering," *London: The Institution of Mining and Metallurgy.*

Kelly, J.M.H. and Martin, P.L. (1986). "Construction Works on or Near Landslides," *In Proc. Symposium of Landslides in South Wales Coalfield.* Polytechnic of Wales, pp. 85-103.

Kennedy, B.A. and Niermeyer, K.E. (1970). "Slope Monitoring System Used in the Prediction of a Major Slope Failure at the Chuquicamata Mine, Chile," *In Proc. Symp. on Planning Open Pit Mines,* Johannesburg, pp. 215-225.

LCPC: Laboratoire Central des Ponts et Chaussées (1994). "Surveillance Des Pentes Instables," *Guide Technique*, Paris.

Lee, E.M., Moore, R., Burt, N., and Brunsden, D. (1991). "Strategies for Managing the Landslide Complex at Ventnor, Isle of Wight," *Proc. Int. Conf. on Slope Stability Engineering, Isle of Wight*, Thomas Telford, London, pp. 219-225.

Leroi, E. (1996). "Landslide Hazard – Risk Maps at Different Scales: Objectives, Tools and Developments," *In Landslides, Proc. Int. Symp. on Landslides,* Trondheim, 17-21 June (Ed. K. Senneset), pp.35-52.

Malone, A.W. (1998). "Risk Management and Slope Safety in Hong Kong," *In Slope Engineering in Hong Kong*, J.N. Kay and K.K.S. Ho (Eds.), Rotterdam, A.A.Balkema, pp.3-17.

Mikkelsen, P.E.(1996) "Field instrumentation". In: A.K. Turner and R.L. Schuster, op. cit., pp. 278-314.

Popescu, M.E. (1984). "Landslides in Overconsolidated Clays as Encountered in Eastern Europe," State-of-the-Art Report," *In Proc. 4th Intern. Symp. on Landslides*, Toronto, pp. 83-106.

Popescu, M.E. (1996). "From Landslide Causes To Landslide Remediation, Special Lecture," *In Proc. 7th Int. Symp. on Landslides*, Trondheim, 1:75-96.

Popescu M.E. (2001). "A Suggested Method for Reporting Landslide Remedial Measures," *International Association of Engineering Geology (IAEG) Bulletin*, 60, 1:69-74

Popescu, M.E., and Seve, G. (2001). " Landslide Remediation Options after the International Decade For Natural Disaster Reduction (1990-2000), Keynote Lecture", *Proc. Conf. Transition from Slide to Flow - Mechanisms and Remedial Measures*, ISSMGE TC-11, Trabzon, pp. 73-102

Powell, G.E. (1995). "Recent Changes in the Approach to Landslip Preventive Works In Hong Kong," *In Proc. 6th Intern. Symp. on Landslides*, Christchurch, 3:1789-1795.

"Primer on Natural Hazard Management in Integrated Regional Development Planning (1991)," *Department of Regional Development and Environment Executive Secretariat for Economic and Social Affairs Organization of American States*, Washington, D.C.

Saito, M. (1965). "Forecasting the time of occurrence of slope failure," *In Proceedings of 6th International Congress of Soil Mechanics and Foundation Engineering*, Montreal, v. 2, pp. 537-541.

Sassa, K. (1996). "Prediction of earthquake induced landslides," *In Proceedings of the 7th International Symposium on Landslides*, Trondheim, Norway, v.1, pp. 115-132.

Soeters, R. and van Westen, C.J. (1996). "Slope Instability Recognition, Analysis, and Zonation," *In Landslides: Investigations and Mitigation*, Special Report 247, Transportation Research Board, National Research Council, Washington, D.C., pp. 129-177.

Schuster, R.L. (1995). "Recent Advances in Slope Stabilization," *Keynote paper. Proc. 6th Intern. Symp.on Landslides*, Christchurch, 3:1715-1746.

Schuster, R.L.(1996). "Socioeconomic Significance of Landslides," Chapter 2 in Landslides – Investigation and Mitigation, Special Report 247, Transportation Research Board, National Research Council, Turner, A.K. and Schuster, R.L., (eds.), National Academy Press, Washington, D.C., pp.12-31.

Skempton, A.W., Leadbeater, A.D., and Chandler, R.J. (1989). "The Mam Tor Landslide, North Derbyshire." *Phil. Transactions of the Royal Society*, London, A329: 503-547.

Smith, D.I. (1984). "The Landslips of the Scottish Highlands in Relation to Major Engineering Projects." *British Geological Survey Project 09/LS. Unpublished report for the Department of the Environment.*

Terzaghi, K. (1950). "Mechanisms of Landslides," *An Application of Geology to Engineering Practice – Berkey Volume, Geological Society of America*, Berkley, pp. 83-123.

United Nations (1996). "Mudflows - Experience and Lessons Learned from the Management of Major Disasters," *United Nations Department of Humanitarian Affairs*, New York, 139 p.

Varnes, D.J. (1978). "Slope Movements and Types and Processes," *In Landslides Analysis and Control,* Transportation Research Board Special Report, 176:11-33.

Varnes, D.J. and The International Association of Engineering Geology Commission on Landslides and other Mass Movements (1984). "Landslide Hazard Zonation: A Review of Principles and Practice," *Natural Hazards*, v.3, Paris, France. UNESCO, 63p.

Voight, B., and Kennedy, B.A.(1979). "Slope failure of 1967-1969, Chuquicamata Mine, Chile: In Rockslides and Avalanches," Elsevier, Amsterdam, v. 2, pp. 595-632.

Wieczorek, G.F. 1984. Preparing a detailed landslide-inventory map for hazard evaluation and reduction. Bull. Assoc. Engng. Geol. 21:337-342.

Wilson, R.C., Wieczorek, G.F., (1995). "Rainfall Thresholds for the Initiation of Debris Flows at La Honda, California, Environ. Eng. Geosci. 1:11-27.

Working Party on World Landslide Inventory (WP/WLI): International Geotechnical Societies' UUESCO Working Party on World Landslide Inventory, Cruden, D.M., Chairman, A Suggested Method for a Landslide Summary, *International Association of Engineering Geology Bulletin* (IAEG), 43: 101-110, 191.

Working Party on World Landslide Inventory (WP/WLI): International Geotechnical Societies' UNESCO Working Party on World Landslide Inventory. Working Group on Landslide Causes - Popescu, M.E., Chairman. (1994). A Suggested Method for Reporting Landslide Causes, *International Association of Engineering Geology Bulletin IAEG*, 50:71-74.

Wu, T.H. (1991). "Soil Stabilization Using Vegetation," *In Proc. Workshop on Biotechnical Stabilization*, University of Michigan, Ann Arbor, A-1 to A-32.

Zaruba, Q., and Mencl, V. (1982). "Landslides and Their Control." Elsevier, Amsterdam, 324 p.

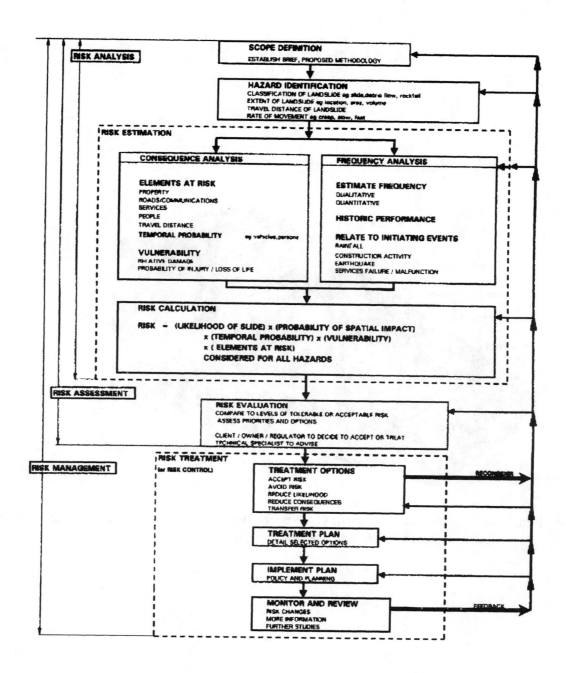

Figure1.   Process   of   landslide   risk   management   (from   Australian   Geomechanics Society, 2001).

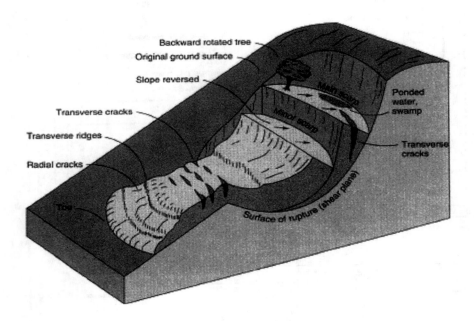

Figure 2.   Nomenclature of various features of a complex earth – slide flow (from Cruden and Varnes, 1996).

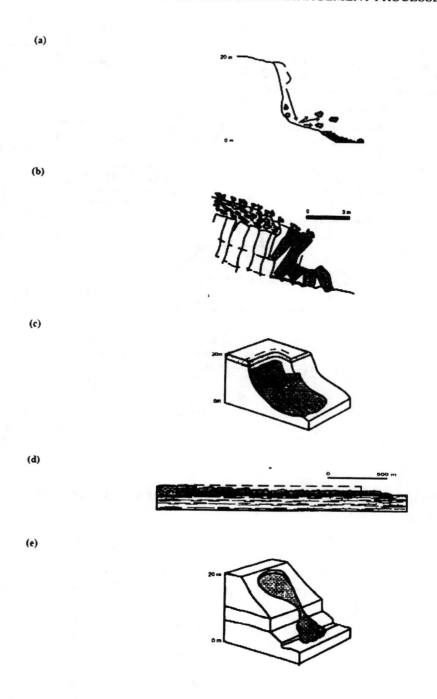

Figure 3.   Types of movement: *(a) fall, (b) topple, (c) slide, (d) spread, (e) flow.* Broken lines indicate original ground surfaces; arrows show portions of trajectories of individual particles of displaced mass (from Cruden, 1991).

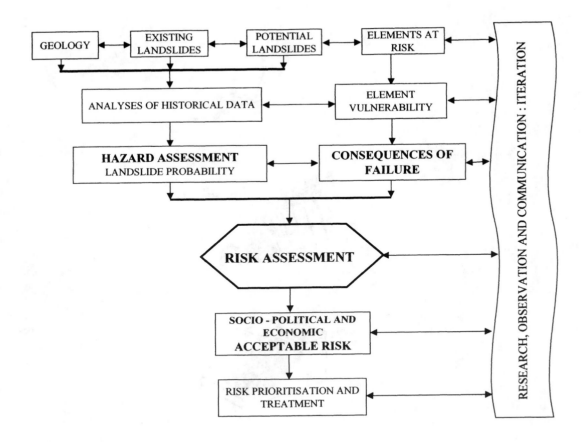

Figure 4.   Methodology for landslide risk assessment (from Chowdhury et al, 2001).

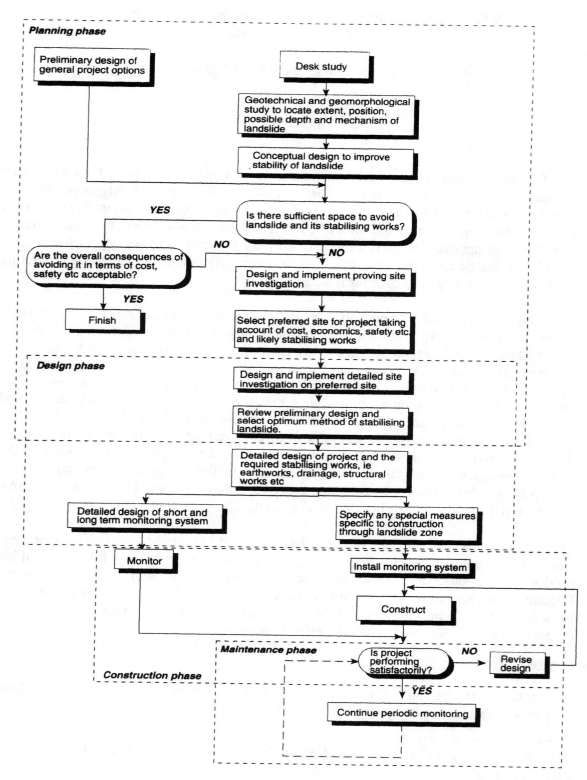

Figure 5.   Various phases involved in planning, design and construction of landslide remedial works (from Kelly and Martin, 1986).

A Preliminary Study of Geologic Hazards for the Columbia River Transportation Corridor

Yumei Wang[1] and Amar Chaker[2]

ABSTRACT

The Columbia River Transportation Corridor is a significant east-west transportation corridor for the Pacific Northwest that includes U.S. Interstate Highway 84, two transcontinental rail lines, and inland water navigation on the Columbia River. Major electric power and gas lines and lines of communication are also located on that corridor. The stretch of approximately 150 miles of corridor covered in this study includes three large hydroelectric facilities, each of which has over 1,000 megawatt electrical generating capacity and a navigational lock for river commerce.

The study region includes diverse geology settings ranging from flat river valleys to steep gorge slopes and is exposed to multiple natural hazards. It has experienced damaging earthquakes, volcanic eruptions, heavy winter storms, and destructive landslides. The objective of the study is to examine the complex relations among different modes of transportation and geologic hazards, and to assess their importance for the community and the region. Using data on the engineered systems (bridges, roads, dams, and railroads), data on hazards (geologic hazards such as rock fall landslides, debris flow landslides, volcanic landslides, earthquake ground shaking, floods, and dam stability-related hazards) and limited economic and commerce data, a preliminary review of the interdependencies and of the overall vulnerability in the region was undertaken. Two low-probability worst-case scenarios that could occur between near Hermiston in Morrow County and the Portland area have been identified.

The study results indicate that geologic hazards in the Columbia River Transportation Corridor can have a severe, long lasting impact on the economy of Oregon, affect productive capacity, and slow the pace of economic growth and development. Recommendations are made for additional studies to be conducted to better evaluate the risks and for mitigation measures to be implemented to lower the vulnerability of this transportation corridor that is vital to Oregon's economy.

INTRODUCTION

The multi-modal Columbia River Transportation Corridor stretches for 150 miles along Oregon's Interstate Highway 84 from near Hermiston in Morrow County on the east to

---

[1] Oregon Department of Geology and Mineral Industries, 800 NE Oregon St., #28, Portland, OR 97232

[2] Civil Engineering Research Foundation, 1801 Alexander Bell Drive, Reston, VA 20191-4400

Portland on the west (Figure 1). It includes a major Interstate Highway, I-84, two transcontinental rail lines, Union Pacific (UP) and Burlington Northern Santa Fe (BNSF), the Columbia River inland water navigation, major electric power and gas lines, and telecommunication lines. The main terminus of the corridor is Portland. The Portland International Airport and trucking also play major roles in the transportation system of the region.

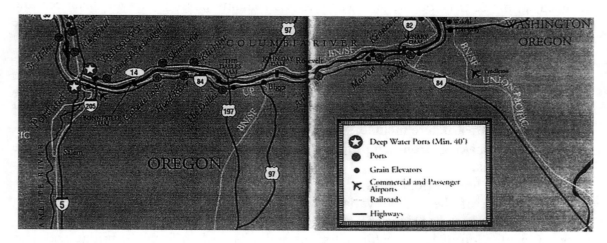

Figure 1. Map of the greater study area showing the Columbia River Transportation Corridor along the Oregon-Washington border (The Great Waterway, 1998).

The study region is characterized by extreme climatic conditions ranging from heavy rainfall to high desert aridity. It includes diverse geology settings ranging from flat river valleys to steep gorge slopes. Damaging earthquakes, volcanic eruptions, heavy winter storms, and destructive landslides have been experienced in this region. The study area also includes stretches of the Columbia River, a major river and dam system that is critical for flood control and power generation. It is located in the lower reaches of a large watershed, which includes part of the Province of British Columbia and the States of Washington, Oregon, Wyoming, Nevada, Idaho, Montana, and Utah.

NATURAL DISASTERS AND TRANSPORTATION SYSTEMS

Transportation systems provide an essential function to society by allowing the movement of people and goods and are the key to commerce and economic activity. They provide the mobility that allows us to conduct business, commute, travel, and enjoy recreation. They are lifelines without which citizens and businesses could not function. Transportation represents a substantial share of a country's gross domestic product (11% for the United States).

Geologic hazards such as major earthquakes or landslides can develop into natural disasters with far-reaching impacts on communities, and also on the transportation infrastructure. The failures of the Cypress Viaduct and the Oakland- San Francisco Bay Bridge during the 1989 Loma Prieta earthquake, of major freeway interchanges during the 1994 Northridge earthquake, and of the Hanshin Expressway during the 1995 Kobe earthquake provide striking examples of damage to the transportation infrastructure. For earthquakes, the causes of such disruptions are many, among them bridge collapse, landslide, impending collapse of an adjacent structure, bursting of a nearby water or natural gas pipe, settlement or compaction, liquefaction, lateral spreading, surface rupture, rock falls, and more. In these instances, one or more links or nodes of the transportation network are rendered unusable, the transportation system's capacity is reduced, gridlock may occur, and when access remains possible, travel distances and times are greatly increased. Damage to one critical link can have impacts far beyond the local area.

Our current understanding of the seismic response of structures and their modes of failure are based on the principles of structural dynamics, the results of recent research and the findings of post-earthquake investigations. As examples, possible causes for failure of transportation infrastructure include:

- Severe shaking from amplification of the ground motion due to local site conditions.
- Fault rupture at bridge sites
- Influence of the spatial variation of ground motion on the response of long structures such as bridges.
- Liquefaction of loose, saturated sands and silts – often found at bridge and dam sites along rivers.
- Settlement of the abutment fill material, possibly with slumping and abutment rotation.
- Pounding between adjacent structures coming in contact during earthquake shaking because they are too close together.

The primary features of transportation networks relevant to their resiliency to disasters include the following:

- They are decentralized and typically extended over wide geographic areas, and so is their vulnerability.
- Failure at one point along a link often means failure of the link.
- Transportation infrastructure frequently follows river valleys, where population centers are located and it is often easier and cheaper to build. However, such location makes the infrastructure prone to flooding and, in seismic areas, susceptible to liquefaction damage.
- Utilities, including pipelines, often follow transportation infrastructure right-of-ways, and cross valleys using bridges of the road system. Due to the physical proximity of these lifeline systems and to their functional interdependency, interaction between them can be significant.

Events of recent years offer striking examples of disasters impacting transportation systems. The earthquakes of Loma Prieta, California (October 17, 1989), Northridge, California (January 17, 1994) and Kobe, Japan (January 17, 1995), dramatically demonstrated the devastating impact earthquakes could have on highway bridges not adequately protected against seismic forces. So did the bridge collapses observed during the 1999 earthquakes in Kocaeli, Turkey and in Chi-Chi, Taiwan.

As the above examples indicate, bridges are among the most vulnerable components of highway systems, which in turn are the backbone of the transportation system. While disruptions of the highway system are the most frequent, railroads, ports, terminals, airports, waterways, utilities, and the interaction among these modes can also be affected by earthquakes. The 2001 earthquake at Nisqually, Washington, caused damage to the Seattle airport control tower, which forced closure of the airport. During the same earthquake, liquefaction rendered the pavement of the Boeing airfield and a section of the railroad unusable (Figure 2). The overturned Oran-Algiers train crossing the fault at the time of main shock during the 1980 earthquake at El Asnam, Algeria and the rails of the railroad between Guatemala City and Puerto Barrios bent by the 1976 earthquake in Guatemala are extreme examples of railway system disruptions. Similarly, the port of Kobe, Japan, suffered extensive damage due to liquefaction during the 1995 earthquake.

Figure 2. Damage to railway from liquefaction in the 2001 Nisqually, Washington earthquake (photo credit: William Byers).

Pipelines are not immune to earthquakes either. The motion on the Oued Fodda fault during the 1980 earthquake at El Asnam sheared a major water pipeline. Gas pipelines in the San Francisco Marina district were ruptured during the 1989 Loma Prieta earthquake, causing major fires. During the 1999 earthquake at Chi-Chi, Taiwan, natural gas pipelines were damaged due to liquefaction.

Interaction among different modes of transport and transport modes with utilities can affect the actual vulnerability of transportation systems. Illustrative examples are given below:

- Damage to a highway bridge crossing over a railroad can interrupt railway traffic.
- The capacity of an airport can be reduced because air traffic controllers and passengers have difficulty reaching the airport using surface transportation. In such a case, the highway and air modes of transport are "in series." Airport facilities are examples of inter-modal interfaces, the damage of which can have major consequences far beyond their immediate location.
- In some cases, when different modes of transportation are "in parallel," one mode with little or no damage (e.g., transit) may have to assist another mode (e.g., highway system), which may have suffered extensive damage, and then function only in degraded fashion.
- Utility lines, collocated or in the vicinity of transportation infrastructure, can affect it as well. For example, a water main break, a gas pipe leak or downed power lines can all lead to road closures. Loss of traffic signals can impair the operation of the transportation infrastructure and hamper relief efforts.

It should be realized, however, that disruptions of the transportation systems do not always lead to a complete breakdown. To some extent, transportation networks can adapt to the loss of a link or node. In some cases, the flow capacity is reduced, but access remains possible (Figure 3). Unless a link is the only way to access a given area, the redundancy usually present in transportation networks provides alternate routes. In some cases, it may be possible to rely on the substitution of one mode for another. Disruptions of components of the transportation systems do not always constitute an on-or-off situation, and it is often possible for the system as a whole to continue to function at a reduced capacity or in a degraded mode (e.g., reduced capacity, longer travel distances and times).

Figure 3. One of two parallel bridges failed in the 2003 Zemmouri, Algeria earthquake causing traffic to be rerouted to the undamaged bridge (Photo credit: Mark Yashinsky).

The devastating effects of disasters occur usually in the built infrastructure itself, with damage to nodes and links of the transportation network, and to the utilities located in the right-of-way. The direct loss (cost of restoring the damaged node or link) is compounded by additional indirect losses. Since transportation systems are critical to disaster response and recovery, delays in the arrival of emergency vehicles and the evacuation of casualties may result in increased fatalities. A transportation system functioning in degraded mode impedes the arrival of food, supplies, and heavy equipment on site and thus delays recovery. While the transportation system is functioning in degraded mode, additional losses are incurred because, as travel distances and times are lengthened, debris-removal, reconstruction, repair, and retrofitting activities are slowed down. In some cases the maximum transportable load is reduced. Such disruptions in transportation systems can have a severe impact on the economy of a region for months or even years, affect productive capacity, and slow the pace of economic recovery.

In view of the critical role of transportation systems in emergency response and disaster recovery and of the large direct and indirect losses that can be incurred, it is essential that the vulnerability of the transportation infrastructure be understood.

THE MULTI-MODAL COLUMBIA RIVER TRANSPORTATION CORRIDOR

Statewide in Oregon, the largest percentage of tonnage is moved by truck, rail, and barge, respectively. The statewide freight movements by modes for tonnage and dollar value are shown in Table 1. The top five commodities shipped to, from, and within Oregon by all modes in 1998 are shown in Table 2.

Table 1. Freight shipments to, from and within Oregon in 1998 (US DOT)

Mode	Weight (Million tons)	Value ($ Billions)
Highway	220	165
Rail	53	18
Water	16	3
Total	289	186

Table 2. Commodities shipped to, from and within Oregon in 1998 (US DOT)

Commodity	Weight (Million Tons)	Value ($ Billions)
Lumber/Wood Products	105	41
Farm Products	38	39
Secondary Traffic	38	23
Freight All Kinds*	18	16
Clay/Concrete/Glass/Stone	18	13
Total	217	132

*All Kinds refers to "Single freight, which is charged irrespective of the commodity" (www.eyefortransport.com/glossary/ab.shtml).

Some of the modal share in the Columbia River Transportation Corridor is "inter-modal," which refers to connecting different modes of transportation and/or transferring freight from one mode to another at facilities such as ports, terminals, stations, or airports. For example, a portion of lumber harvested in Oregon and bound for inland markets by rail travels to the rail reload facilities by truck. Grain and potash rail shipments, on the other hand, travel to the Port of Portland for export by ship. Freight movements can also take place at non-inter-modal facilities, such as distribution centers and warehouses, manufacturing plants, truck reload facilities, and terminals (ODOT, website).

The Portland region has a high concentration of freight activity that includes inter-modal operations at ports, rail yards (or terminals), trucking and industrial yard facilities. The Port of Portland, Port of Vancouver, and other private Portland-Vancouver regional ports handle most of the barged goods. However, some barged freight, such as corn, passes through the Columbia River Gorge and Portland-Vancouver region on rail without stopping to Kalama, downstream of Portland.

A considerable amount of "through freight" passes through the Portland-Vancouver region, linking, e.g., Puget Sound and inland markets. Estimates indicate that more than 15 million tons of rail freight moved between markets east and north of the Portland-Vancouver region without stopping (Scott Drumm, oral communication, December 10, 2003).

In the Columbia River Transportation Corridor, trucking is the most used transportation mode, followed by rail, inter-modal transport, then barge. Tables 3 and 4 provide the top five commodities that are transported through the corridor both to and from the Portland-Vancouver metropolitan area, including all modes of transportation by value and by weight. The data from Tables 3 and 4 are from the 2002 "Commodity Flow Forecast Update and Lower Columbia River Waterborne Cargo Forecast" report by DRI-WEFA, Cambridge Systematics, and BTS Associates.

Table 3. Top 5 Commodities by Value (Columbia River Transportation Corridor)

Commodities	Short Tons (in thousands)	$ Value (in millions)
Vehicles	1,960	$24,053
Textiles, leather, and articles	569	$9,1612
Foodstuffs and alcoholic beverages	5,332	$7,7835
Cereal grains	10,870	$6,649
Mixed freight	2,770	$5,506

Table 4. Top 5 Commodities by Weight (Columbia River Transportation Corridor)

Commodities	Short Tons (in thousands)	$ Value (in millions)
Cereal grains	10,870	$6,649
Gas, fuel, petroleum/coal products	5,711	$1,928
Base chemical	5,471	$3,744
Foodstuffs and alcoholic beverages	5,332	$7,785
Wood products	3,621	$1,240

The entire Columbia River system carries about $14 billion worth of goods each year, provides the means to ship goods to over 1,000 companies in Portland, and, ranks fourth in the nation and first in the West Coast in agricultural export tonnage (Port of Portland website). Oregon's ports near the study region include Umatilla, Morrow, Arlington, The Dalles, Hood River, Cascade Locks, and the Port of Portland. Grain elevators are located near Umatilla, Morrow, Arlington, Biggs, The Dalles and in Portland.

Three major hydroelectric facilities exist in this study region. Each of the hydroelectric facility consists of a dam, powerhouse, spillway, navigational lock, and fish passages. Bonneville Lock and Dam, which is located 40 miles east of Portland at the head of tidal influence from the Pacific Ocean, has a 1,059-MW electricity generating capacity. The Dalles and John Day facilities upriver have electricity generating capacities of 1,636 MW and 2,160 MW, respectively. Barge traffic through Bonneville Lock moved 9.4 million tons in 1996, which is equivalent to 940 100-car unit trains or over 180,000 trucks. Predominant exports include wheat, soda ash, potash and hay to Japan, South Korea, Brazil, and Taiwan. Major imports are automobiles, petroleum products, steel, and limestone from Japan, South Korea, China, and Australia.

The Port of Portland, Port of Vancouver and smaller private ports handle some of the barged commodities. In 2000, the Port of Portland handled over $10 billion in imports and exports (Port of Portland website, 2003). According to the commodity flow study on 1997 data, about $4 billion was from barged commodities (Scott Drumm, Port of Portland, oral communication, December 2003). The waterborne commodity flow handled by the Port of Portland alone for 2000 is shown on Table 5.

Table 5 Waterborne Commodity Flow by Port of Portland
Cargo (2000 actual volumes)

Containers (Twenty-foot equivalent units)	290,000
Breakbulk (Metric Tons)	385,000
Automobiles (units)	585,000
Bulk Grains (Metric Tons)	2,919,000
Bulk Minerals (Metric Tons)	3,827,000

Note: Breakbulk refers to specific commodities (other than automobiles, bulk grains and bulk minerals) that are not containerized, such as long or heavy objects like logs, lumber and stacked steel plates.

The I-84 transportation infrastructure is dependent on roads, bridges, tunnels, and clear passages to transportation hubs, such as Portland International Airport and inter-modal industrial parks. Many portions of the roads and rails are subject to landslides and, in places, overpass collapses. I-84 has approximately 130 bridges between Portland and Hermiston.

The rail infrastructure is dependent on railways, bridges, tunnels, and facilities. Much of the rail freight terminating in Portland originates in Washington, California, Idaho, Montana, Wyoming and Illinois. Portland is at the western end of both the UP and BNSF railway lines through the Columbia River Gorge. The two railroads moved a combined 128 million gross tons over their lines in the Gorge in 1999—67.5 million on UP and 61 million on BNSF. These main lines are the most heavily used rail system in the Pacific Northwest. The Union Pacific terminal is a collection of facilities provided by a railway at a terminus and/or intermediate point for freight and the receiving, classifying, assembling, and dispatching of trains. UP has major terminal facilities at Brooklyn Yard in southeast Portland, Albina Yard in northeast Portland and Barnes Yard in north Portland. UP also has inter-modal ramps at both Brooklyn and Albina. The Burlington Northern Santa Fe major facility is located in Vancouver, Washington. In Portland, BNSF terminals include Willbridge in northwest Portland, which is inter-modal, and Rivergate in north Portland, which is the largest receiver and shipper of freight in the region. UP and BNSF jointly own the Portland Terminal Railroad with a major terminal at Lake Yard in northwest Portland. It serves the Port of Portland facilities along the west shore of the Willamette River and industries located in northwest Portland.

GEOLOGIC HAZARDS IN THE STUDY REGION

A report on the geologic history of the Columbia River Gorge was first published in a Geological Society of America field trip guide by Wang and others (2002). The description of the geologic hazards the study region is exposed to are based on that report.

Landslides and Landslide Hazards

The Gorge with its steep slopes and high rainfall commonly experiences slope failures. Active landslides of diverse types can be found, each producing its own specific hazard. For example, fast-moving landslides such as rockfalls, rock avalanches, and debris flows pose direct threats to life and property (Figure 4). Slow-moving landslides that bulge near the toe pose maintenance concerns for highways, railroads, and other lifelines. River bank failures and underwater landslides, including lateral spreading induced from high pore water pressures or ground shaking, can also occur.

The hydrogeologic conditions of the stratigraphic units increase the risk of sliding. Precipitation readily penetrates the columnar and hackly joints of the upper basalt flows and volcaniclastic deposits. This water tends to collect when it intercepts the less permeable Eagle Creek Formation. This condition is repeated where the Eagle Creek Formation, which has a much higher hydraulic conductivity and porosity, is in contact with the Ohanapecosh Formation, which has a very low hydraulic conductivity. Both geologic contacts provide weak, slide-prone surfaces for the thick, dense overlying basaltic lava flows (Waters, 1973).

Although the rocks typically dip only gently toward the south in the western part of the Gorge, the nature of the landslides is largely controlled by this dip. Washington landslides tend to be of a large scale and produce low slope angles after coming to rest. On the Oregon side of the Gorge, although the geology is the same, the rocks are less susceptible to large-scale landsliding because they dip into the Gorge valley walls, but landsliding problems tend to persist. For example, according to the Oregon Department of Transportation, the Fountain Landslide, three miles east of Cascade Locks, has remained active for more than 35 years and regularly causes distress to I-84 (Shuster and Chleborad, 1989).

Figure 4. Steep, landslide-prone slopes, including the Oregon Shore landslide and barge traffic near Cascade Locks, Oregon.

In February 1996, heavy rains triggered several debris flows, including the large Dodson debris flow, which flooded a home (Figure 5), buried U.S. Interstate Highway 84 and the railroad and entered the Columbia River before entering the river.

Figure 5. The 1996 Dodson debris flow flooded this residence before crossing I-84 and the railway and entering the Columbia River (photo: ODOT).

Rapidly moving landslide hazards have been mapped by DOGAMI for the western portion of the transportation corridor, from Hood River County on the east to Portland on the west (Hofmeister and others, 2002, 2003). This portion has historically experienced rapidly moving landslides in the steeply sloped areas; however, other portions of the transportation corridor can experience such landslides as well. The website http://www.coastalatlas.net/learn/topics/hazards/landslides provides transportation data and coverage for topography, orthophotos and shaded elevation (Oregon Ocean-Coastal Management Program). In addition, it includes the DOGAMI landslide hazard maps for the transportation corridor mentioned above. Figure 6 shows custom maps that include the Bonneville Dam and three close-up maps of the same region: landslide hazards on topography, landslide hazards on an orthophoto and landslide hazards on shaded relief.

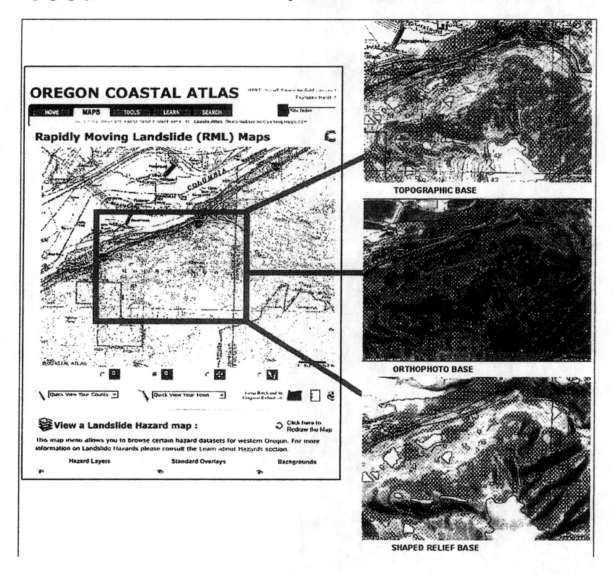

Figure 6. An example of DOGAMI rapidly moving landslide maps on three base maps near Bonneville Dam, located in the upper left part of the insets (modified from Oregon Ocean-Coastal Management Program and Hofmeister, and others, 2002).

Additional digital coverage for this transportation corridor can be obtained at http://www.inforain.org/interactivemapping/gorge.htm. This website allows for interactive custom maps. Landslide maps are also available. For example, the mapped landslides in the central Gorge region are shown on Figure 7.

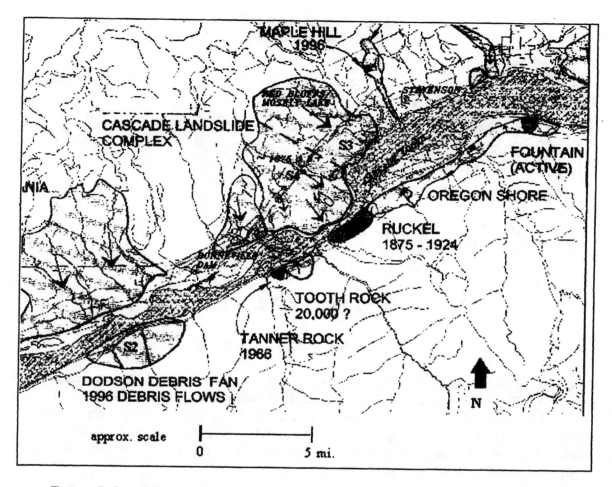

Figure 7. Landslide map of the central Gorge region (Squier Associates, 1999).

The Oregon Department of Transportation has also mapped the locations that are susceptible to rockfall risks (Figure 8). On December 9, 2003, a rockfall was triggered by freeze-thaw exposure at one of their mapped high-risk locations. Rocks estimated at 5 ft in diameter fell from a 150-ft-high vertical slope with overhanging areas and hit I-84 before entering the river. The rockfall damaged several cars and caused a closure of the I-84 for several hours in both directions. Rock falls in the Gorge have been known to have caused fatalities.

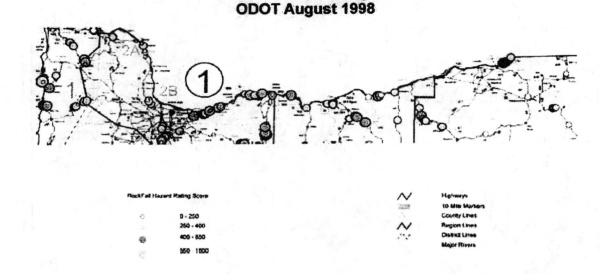

Figure 8. Oregon Department of Transportation rockfall risk map.

Episodic debris-flow activity near Multnomah Falls, Warrendale, and Dodson has forced road and rail line closures. In 1996, the debris flows at Dodson and Tumalt Creek in Warrendale closed I-84 and the railroad for five and three days, respectively. Again in December 2001, a series of debris flows buried and closed the I-84 Exit 35 on-ramp in Dodson for over 12 days. This debris flow was very fluid, and mud lines were observed as high as 23 m (75 ft) up tree trunks in the transport zone.

Earthquakes and Earthquake Hazards

Earthquake hazards are from shallow, crustal earthquakes and Cascadia Subduction Zone earthquakes, which occur at intervals ranging from decades to hundreds of years. The USGS has mapped peak ground accelerations with a two-percent probability of exceedance in 50 years for the entire region (Figure 9). These ground shaking maps, which are available on the USGS website, indicate a significant risk to the dam facilities. The shaking is expected to trigger numerous rockfalls and landslides along the corridor, some of which would impact the waterways. Potentially costly coseismic geohazards include seiches, landslides, lateral spreading, liquefaction, and fault rupture. Volcanic earthquakes are considered to be less likely a risk than earthquakes from a tectonic source.

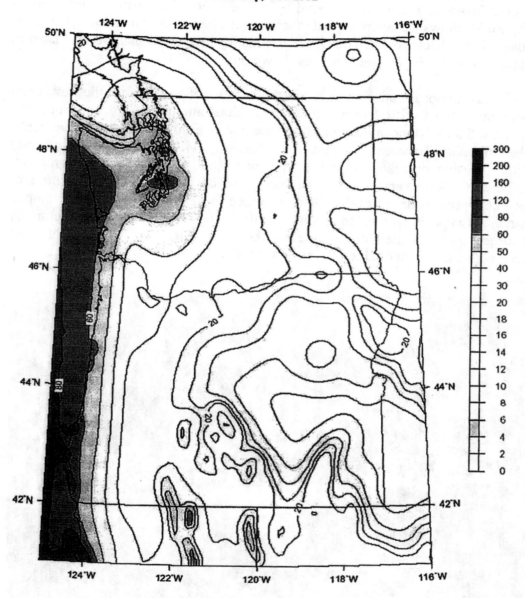

Peak Accel. (%g) with 2% Probability of Exceedance in 50 Years
USGS Map, Oct. 2002

Figure 9. U.S. Geological Survey earthquake shaking map (Frankel and others, 2002).

Earthquake shaking in 2001 from the magnitude 6.8 intraplate earthquake at Nisqually, Washington, triggered rockfalls in the Gorge (David Keefer, U.S. Geological Survey, oral communication, 2001). Past earthquakes are suspected to have triggered large-scale landsliding.

Volcanoes and Volcanic Hazards

Volcanic hazards in the Cascade Range, including eruptions and lahars (fast-moving landslides or debris avalanches are triggered by volcanic activity), occur at intervals of decades. Figure 10 shows the hydro-electric facilities at The Dalles with Mount Hood, an active volcano, in the background. The USGS has mapped the areas that are estimated to be in the lahar risk zones for Mount Hood, as shown on Figure 11.

Mount Hood's last major eruption occurred in the 1790s, not long before the Lewis and Clark expedition to the Pacific Northwest. A tremendous amount of volcanic rock and sand entered the Sandy River drainage at that time and is easily seen on the riverbanks between the Columbia River and U.S. Interstate Highway 84. Other events include one lahar in 1980, causing one fatality, and another one 1,500 years ago with significant deposits. About 100,000 years ago, a large portion of the north flank and summit of Mount Hood collapsed. A debris avalanche was formed from this collapse and developed into a lahar that swept down the Hood River Valley. The lahar crossed the Columbia River and surged up the White Salmon River on the Washington side. The lahar deposit is 400 ft deep where the town of Hood River now stands.

Figure 10. View of The Dalles hydroelectric facilities and Mount Hood, an active volcano, in the background (USACE photo).

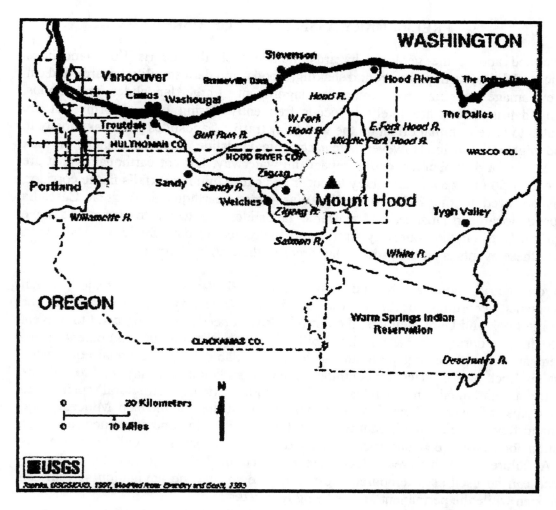

Figure 11. U.S. Geological Survey map showing volcanic lahar hazard zones (Gardner and others, 2000)

Flood Hazards

Before the dam system was constructed in the Columbia River and its tributaries, the flooding hazard was significant. In 1948, the town of Vanport on the Columbia River was destroyed due to the failure of a railroad embankment and ensuing severe flooding. The current system of dams has largely controlled the flooding. However, due to the large watershed, flooding damage is still a hazard to communities and infrastructure. In the 1996 floods, liquefaction was triggered by the excess pore water pressures in the vicinity of the levee that protects the Portland International Airport. Sand boils were observed on the landward side of the levee. If flooding conditions are worse than in 1996, or if earthquake shaking occurs, it is possible that more extensive liquefaction and levee failure could occur. The levee, which was studied by the USACE, was found to be stable for river levels below an elevation of 42.2 ft (USACE, 2001) and potentially unstable for higher river levels. In addition, two worst-case scenarios (discussed below) reveal that flooding in low-lying areas, including parts of downtown Portland and Swan Island are possible.

PRELIMINARY ASSESSMENT OF RISK

As described above, the geologic hazards include rockfalls, debris flow landslides, volcanic landslides, earthquake ground shaking, floods, and dam stability hazards and can produce damage ranging from minor to significant. Large landslides and floods, including debris flows, are likely to occur frequently during heavy storms but are anticipated to have minimal effect on the waterway traffic. Landslides pose risk mostly to rail and highway traffic. However, a catastrophic landslide, such as a lahar or rock avalanche near a dam facility, could impact the waterway. Distant earthquakes that are centered over 50 miles away are likely to occur and cause minor rockfalls in the corridor, such as occurred in the 2001 Nisqually, Washington, earthquake. A great Cascadia earthquake, local earthquakes, and volcanic landslides are likely to occur and have damaging effects on the waterway traffic and hydroelectric dam facilities. The return period of these events is on the order of hundreds to thousands of years.

The hydroelectric facilities, including the locks and dams, require major capital improvements according to the Pacific Northwest Waterways Association. At the Bonneville Lock and Dam facilities, the north lock wall needs to be upgraded to prevent possible failure during an earthquake. Spillway power distribution equipment needs replacement, yet replacement parts are no longer available for the original equipment. The Dalles Lock has experienced continued aging/degradation that may lead to a future outage if not maintained. The John Day facilities require extensive structural repairs at an estimated cost of over $20 million. The lock and dam are founded on Miocene-aged basalts and flow breccia. Significant movement and distress in concrete structures of the navigation lock have been observed. A new downstream lock gate will be needed by 2008. A failure of the upstream lock gate occurred in 2002. Fortunately, a floating bulkhead can be used as a temporary upstream gate so that the navigation system can remain open while the permanent gate is being repaired.

The authors conducted a preliminary analysis of the bridges in seven counties using HAZUS99 (FEMA, 1999) to estimate damage and losses from earthquake shaking. The preliminary results show 31 bridges (located county wide and not just on I-84) with at least moderate damage from 1,000-yr probabilistic ground shaking levels and $24 million of direct losses to bridges (Wang and Chaker, 2004). However, in mid-2003, during the course of this study, the Oregon legislature passed an important transportation bill (House Bill 2041), which is referred to as Oregon Transportation Investment Act III (OTIA III). This Act augments the Statewide Transportation Improvement Program (STIP) with $2.5 billion. Over half of the $2.5 billion OTIA III funds will be spent on replacing and repairing state bridges, starting January 2004. Because all of I-84 is considered to be a critical freight route, all the bridges with weight capacity restrictions and non-seismic-related maintenance issues will be repaired or replaced. The Oregon Department of Transportation (ODOT) owns 130 bridges on I-84 within the study region. ODOT's draft plan is to replace 25 of the bridges and repair seven of the bridges within the next five years in OTIA III phase 2. The bridges that are being replaced or repaired were constructed between 1942 and 1987. Five of the bridges in this corridor are considered to be highly vulnerable to earthquake hazards. After OTIA III is implemented, only three bridges from the "high priority" earthquake-vulnerability list will remain highly vulnerable to earthquake damage. These three I-84 bridges were each constructed in 1963 and are located within the first half-mile to the I-5 interchange (Don Crowne, ODOT, oral communication, December 2003). A list of the 130 bridges, which includes the replacement, repair, and earthquake retrofit status, can be found in Wang and Chaker, (2004). Vulnerable bridges and overpasses will pose a threat. For example, the I-84 interchange at Bridal Veil (exit 28) will still pose a collapse hazard onto I-84. The asymmetric construction of the off ramp and the absence of an on ramp contribute to vulnerabilities relating to torsion (Figure 12). Also, the movable bridges associated with river navigational locks at the hydroelectric facilities are vulnerable to small movements. If, for example, the movable bridge at Bonneville Dam cannot operate, then the locks will also not operate, thus preventing barge traffic flow.

Figure 12. The Bridal Veil exit 28 off-ramp on I-84 is susceptible to earthquake damage.

ODOT relies on emergency routes during disasters, such as Highway 35 in Hood River County and Interstate Highway I-205 that connects to the Portland Airport. The slope adjacent and east of Highway 35, located just south of the I-84 interchange, is highly vulnerable to landslides (Figure 13). This steep slope exposes a fault that juxtaposes river channel deposits with basalts, and seeps that contribute to its instability. In contrast, the I-205 bridges that approach the Portland Airport from I-84 are considered to be less vulnerable to earthquake damage. These were generally constructed in the early 1980s and are not listed on the high-priority list of bridges vulnerable to earthquake damage. However, because numerous bridges cross over I-84 and I-205, especially on the access routes from downtown Portland, it is possible that collapses of overpasses will hamper initial emergency response efforts. Emergency plans that accommodate a reduced traffic capacity would be prudent.

Figure 13. Highway 35 just south of I-84 is susceptible to landslide failure.

Flooding hazards include flooding inundation of facilities including PDX, the rail terminals, the ports, and other low-lying areas. It is also possible for floods to induce liquefaction and failure of dikes, such as the one adjacent to PDX airport.

TWO WORST-CASE SCENARIOS

Two low-probability, worst-case scenarios for this portion of the waterway have been identified as part of this study (Wang and Scofield, 2003).

Worst-Case Scenario One: Large-scale Landslide
One scenario is a natural recurrence of a large-scale failure of the steep canyon walls similar to the Cascade Landslide Complex. We hypothesize that the Cascade Landslide Complex is affected by earthquake shaking (similar to the 1959 Montana earthquake that

caused the Hebgen Lake landslide) or possibly nearby volcanic activity. Figure 14 shows a computer image of the Cascade Landslide Complex looking west. This landslide complex is monitored with instrumentation and is actively moving at a slow rate. Two distinctly separate headscarps on the right side of the image form the top of the complex. The headscarps appear to be divided by the axis of a synform, which indicates past earthquake activity, and a volcanic feature. The Bonneville landslide includes the westernmost headscarp, which is the landslide mass that temporarily blocked the Columbia River (Figure 15). The slide geomorphology and an assessment of known earthquake-triggered landslides suggest that this slide was induced by an earthquake (Wang and Scofield, 2003). We conclude that it is possible for such a significant landslide to recur in this area. The Bonneville facilities are constructed on the western most and youngest slide, which is on the toe of this landslide. The younger Bonneville landslide blocked and diverted the Columbia River south by over one mile. A similar event occurring today would result in complete disruption of transportation through the Gorge and heavy damage to, and perhaps complete destruction of, major facilities upstream including dams, small cities, and industrial sites. The low-lying areas in Portland, including the port facilities and much of downtown Portland, would be heavily damaged (Figures 16 and 17).

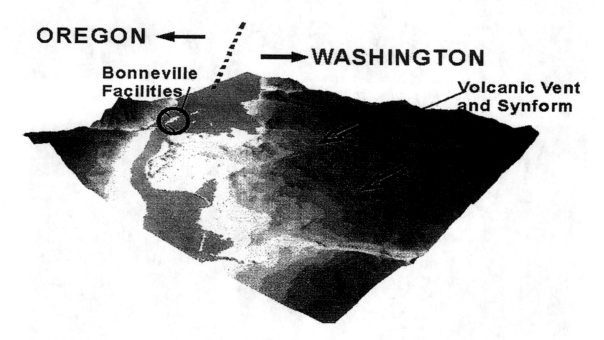

Figure 14. The Cascade landslide complex showing two separate landslides (Image source: R. Wardell and Y. Wang).

Figure 15. Oblique aerial photograph of the Cascade landslide complex and Bonneville Dam facilities (source: D. Cornforth).

Figures 16 and 17. Hypothetical Portland flooding sequence from a Bonneville Dam breach

The Bonneville Lock and Dam includes two powerhouses, a spillway, a navigation lock, fish facilities, and the Bonneville Power Administration (BPA) electrical switchyards. If destroyed, the facility would probably require over 10 years and $2 billion to replace today. It is also possible that only portions of the facility may become inoperable due to geologic hazards. For example, earthquake shaking could lock up the movable bridge that operates in conjunction with the navigational lock. That would render both the bridge and locks inoperable (Figure 18).

Figure 18. Bonneville navigation lock and moveable bridge (source: US ACE Digital Visual Library).

Worst-Case Scenario Two: Flooding

The second scenario is a catastrophic failure and release from John Day Lock and Dam. Such a release has a very low probability and would require an extreme or infrequent event, such as a strong earthquake on a nearby fault. Figure 19 shows that a nearby active thrust fault exists just north of the dam facilities (Bela, 1982). Hartshorn and others (2003) provide seismicity records of the area. This region is in an active compressive setting (although with very slow strain rates) and is therefore capable of generating damaging but infrequent earthquakes. Significant damages would extend downstream to the Pacific Ocean. Damages due to overtopping of dikes (US ACE, 2001) and levees would cause disruption to cities, including portions of downtown Portland, the Portland International Airport facilities, other smaller cities and power generation facilities, and transportation infrastructure. Figure 20 shows the Portland area flood inundation map, where major portions of downtown, the waterfront, and low-lying areas near the Columbia and Willamette Rivers would be flooded.

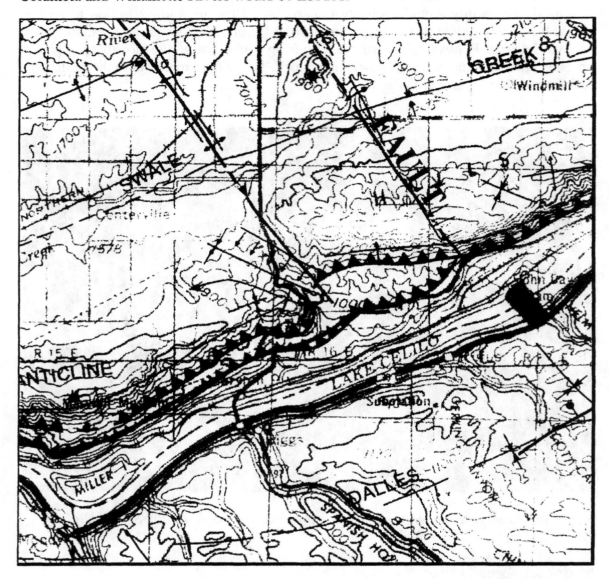

Figure 19. Neotectonics in the John Day Lock and Dam area showing nearby active fault structures. The dam is located to the right of the center (source: Bela, 1982).

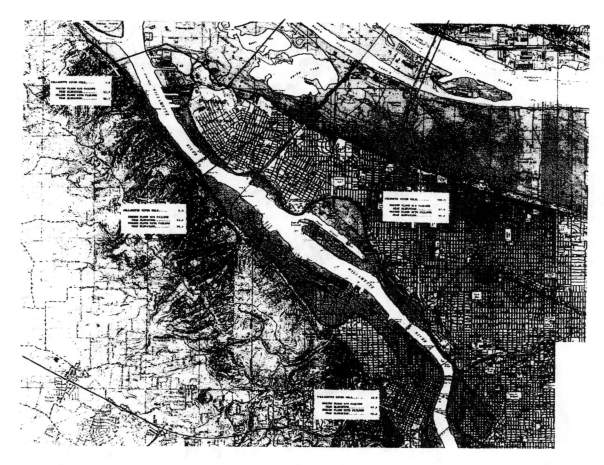

Figure 20. Portland area inundation map from a John Day Dam breach (USACE, 1989).

Inundation from flooding would have an impact on the transportation system and other lifelines, including the waterway traffic and hydroelectric facilities. The highways, including bridges and collocated lifelines, would sustain damage. The low-lying areas, including the port facilities, the rail terminals in Portland, Swan Island, and numerous petroleum tanks located near the Willamette River in northwest Portland, would be flooded. Figure 21 shows the following facilities in the flood inundation zone: the five Port of Portland terminals, the Union Pacific Albina yard and Brooklyn yard, the Burlington Northern Santa Fe Willbridge yard, and the Portland International Airport. The Rivergate Industrial district, Mocks Landing Industrial Park, Swan Island Industrial Park, and Portland International Center are also in the flooding zone.

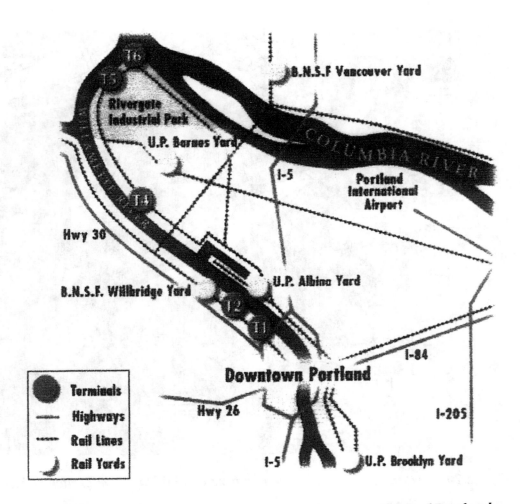

Figure 21. Sketch map showing port and rail terminal locations (Port of Portland website).

CONCLUSIONS AND RECOMMENDATIONS

The Columbia River Transportation Corridor is clearly vital to the economy of the Portland area, of the State of Oregon, and of the Pacific Northwest region. The multiple hazards it is exposed to could cause major damage to its transportation infrastructure and to the major lifelines that run parallel to it. The limited redundancy imposed by the topology of the corridor and the possible interaction among the transportation lines and other lifelines make it particularly vulnerable. The two extreme case scenarios considered in this preliminary study (large scale landslides of the Cascade landslide complex, and flooding following breach of the John Day Lock and Dam) would have severe consequences for the infrastructure located in the corridor and Portland area, and for the population and the economy of the region.

It is recommended that more detailed studies involving key stakeholders be conducted on the worst-case scenarios identified. Research and development that have the potential for improving the disaster resiliency of the Columbia River waterway and the parallel transportation system railways and I-84 should be conducted. Such studies have been identified as needed by the Governor's Oregon Seismic Safety Policy Advisory Commission (OSSPAC, 2000).

For the transportation infrastructure in the corridor, the importance of each component depends primarily on the consequences of an eventual failure. Thus, pertinent questions that may be researched are the following:
- How critical is the component to the function of the overall transportation system?
- Does the component provide access to an essential facility?
- Is the component part of an emergency transportation corridor?
- Is the transportation network topology such that a failure of the component will prevent access to certain areas of the community or of the region?
- Is the transportation network topology such that a failure of the component will have high societal impact for a long duration?
- How will the operations side (i.e., transportation operators, vehicles, traffic safety, power, command control and communications centers, and maintenance) of the transportation system function?

Further studies should address emergency preparedness, hazard evaluation, risk assessment, vulnerability reduction, mitigation plans and implementation of mitigation. For example, table top exercises involving an earthquake and including analyses of infrastructure damage and loss could be conducted with stakeholders and communities to develop a response plan.

ACKNOWLEDGMENTS

The authors would like to thank Janet Benini (now Whitehouse Homeland Security Council), Dolph Diemont and Tom Sachs of the U.S. Department of Transportation for their leadership in supporting this study. We gratefully acknowledge our editor, Craig

Taylor and early reviews by David Scofield and Vicki McConnell. Also, we thank staff at the Oregon Department of Transportation, including Michael Long, Mark Hirota (now with Parsons Brinckerhoff Quade & Douglas), Rose Gentry, Rich Watanabe, Steven R. Kale, Ed Immel, Don Crowne, Brad Dehart, and the U.S. Army Corps of Engineers, including Rich Hannan, James Griffins and Carolyn Flaherty. Scott Drumm of the Port of Portland generously provided assistance with the economic transportation data. Robert Wardwell of the U.S. Geological Survey contributed his GIS and geologic expertise.

REFERENCES

Bela, J.L., compiler, 1982, Geologic and Neotectonic Evaluation of North-Central Oregon, The Dalles 1° x 2° Quadrangle: Oregon Department of Geology and Mineral Industries Geological Map Series GMS-27.

DRI-WEFA, Cambridge Systematics, and BTS Associates, 2002 Commodity Flow Forecast Update and Lower Columbia River Waterborne Cargo Forecast.

Federal Emergency Management Agency, 1999, *HAZUS99 Technical Manual*, Volumes I, II, and III, Washington, D. C., prepared by the National Institute of Building Sciences for the Federal Emergency Management Agency (FEMA).

Frankel, A.D., Petersen, M.D, Mueller, C.S, Haller, K.M, Wheeler, R.L, Leyendecker, E.V, Wesson, R.L, Harmsen, S.C, Cramer, C.H, Perkins, D.M. and Rukstales, K.S, 2002: USGS earthquake shaking Documentation for the 2002 (Update of the National Seismic Hazard Maps), U.S. Geological Survey Open-File Report 02-420.

Gardner, C.A., Scott, W.E., Major, J.J. and Pierson, T.C., 2000, USGS Fact Sheet 060-00, *in* Online Geologic Publications of the Western United States [online]. [Cited 19 November 2003]. <http://geopubs.wr.usgs.gov/fact-sheet/fs060-00/>.

Hartshorn, D.C., Reidel, S.P., and Rohay, A.C., 2003, Third Quarter Hanford seismic report for fiscal year 2003, PNNL-11557-23, Pacific Northwest Laboratory, Richland, Washington.

Hofmeister, R.J., Miller, D.J., Mills, K.A., Hinkle, J.C. and Beier, A.E., 2002, Hazard map of potential rapidly moving landslides in western Oregon [CD-ROM]: Oregon Department of Geology and Mineral Industries Interpretive Map Series IMS-22.

Oregon Ocean-Coastal Management Program,
http://www.coastalatlas.net/learn/topics/hazards/landslides

Oregon Department of Transportation website. <http://www.odot.state.or.us/home/>.

Oregon Seismic Safety Policy Advisory Commission, Oregon At Risk, 2000: Oregon Seismic Safety Policy Commission. <http://www.wsspc.org/links/osspac/oratrisk.pdf>.

Port of Portland, www.portofportland.com

Squier Associates, 1999, Investigation of the Maple Hill Landslide, Stevenson, Skamania County, WA: Report prepared for Skamania County.

The Great Waterway, 1998, Portland Oregon, p. 34-35.

U.S. Army Corps of Engineers, 2001, Seismic Performance of the Columbia River Levee along NE Marine Drive, Portland, Oregon.

U.S. Army Corps of Engineers, 1989, Guidelines for Flood Emergency Plans with Inundation Maps, John Day Project, Columbia River, Oregon and Washington.

U.S. Army Corps of Engineers, Corps of Engineers Digital Visual Library. <http://images.usace.army.mil/main.html>.

U.S. Department of Transportation, Federal Highway Administration. <www.ops.fhwa.dot.gov>.

Wang, Y., Chaker, A., 2004, Geologic Hazard Study for the Columbia River Transportation Corridor, Department of Geology and Mineral Industries, Open File Report OFR O-04-08.

Wang, Y., and Scofield D.H., 2003, Columbia River Waterway Risks, *in* Proceedings of American Society of Civil Engineers Coastal Structures conference, Portland, Oregon, 13 p.

Wang, Y., Hofmeister, R.J., McConnell, V., Burns, S., Pringle, P., and Peterson G., 2002, Columbia River Gorge Landslides, Geological Society of America: Oregon Department of Geology and Mineral Industries Special Paper 36.

Wang, Y., 1999, Risk assessment and risk management in Oregon, American Society of Civil Engineers Technical Council on Lifeline Earthquake Engineering for the 5th U.S. Conference in Seattle, Washington, 10 p.

Waters, A.C., 1973, The Columbia River Gorge: Basalt stratigraphy, ancient lava dams, and landslide dams, in Beaulieu, J.D., ed., Geologic field trips in northern Oregon and southern Washington: Oregon Department of Geology and Mineral Industries Bulletin 77, p.133-162.

Multihazard Mitigation Los Angeles Water System
A Historical Perspective
Le Val Lund[1] and Craig Davis[2]

ABSTRACT

The Water System of the Los Angeles Department of Water and Power (LADWP) has been implementing multihazard mitigations since the early 1900's. The Water System was established in 1902 and shortly thereafter William Mulholland, the first Chief Engineer and General Manager, recognized the need for additional water supply for a growing city. The Los Angeles Aqueduct (LAA) was completed in 1913 to bring water from the eastern Sierra Nevada snowmelt. The LAA brought forth the first major hazard mitigation program; a water supply to help a semi-arid region sustain drought conditions, which incorporated the construction of large storage reservoirs south of the San Andreas fault crossing, along the LAA and in the city, to provide local water storage in the event of a large earthquake or other disasters. The LAA also provided the means for City expansion, and with it multihazard mitigations have been ongoing with the development of the Los Angeles Water System. Most hazard mitigations described herein are related to earthquakes and as such most hazard improvements were recognized and undertaken following the 1971 San Fernando Earthquake as a result of the damage incurred to the Los Angeles Water System. In addition, many system improvements implemented for one hazard help mitigate effects of other hazards and decisions on selecting mitigation alternatives are generally determined through considerations of several hazards. This report discusses the multihazard mitigations implemented for the Los Angeles Water System since the early 1900's, provides an overview of aspects that developed the system's resiliency enabling it to provide service following the 1971 and 1994 earthquake disasters, and gives some perspectives on current seismic improvements being implemented in conjunction with necessary system changes for improving water quality. In addition, interrelationships between mitigations for multiple hazards and conditions where multiple hazards combine to create greater disasters are described. The modern developed LADWP Water System hazard mitigation history provides insight and information useful to others who are interested in improving the resistance of water and other lifeline systems to natural and manmade disasters.

---

[1] Civil Engineer, Los Angeles, CA
[2] Waterworks Engineer, Los Angeles Department of Water and Power

INTRODUCTION

The Los Angeles Department of Water and Power (LADWP) is the largest municipally owned utility in the United States. It exists as an independent proprietary department by virtue of the City Charter of the City of Los Angeles. The LADWP provides water and electric service to more than 3.8 million Los Angeles City residents and businesses in a 465-square mile (120,435 ha) area. The LADWP's operations are financed solely by the sale of water and electricity. Capital funds are raised by the sale of revenue bonds. No tax support is received. A five-member citizen Board of Water and Power Commissioners establishes LADWP policy.

The Water System was created over 100 years ago, in 1902, when the City purchased the private Los Angeles City Water Company. William Mulholland, shown in Figure 1, became the first Chief Engineer and General Manager for the newly created Bureau of Waterworks and Supply, which later became and will be referred to herein as the LADWP, and began enlarging the capacity of the Los Angeles River System. Growth of the Water System is synonymous with that of Los Angeles. As the City grew in the early 20th century, Mulholland saw the need for additional water supply and developed the Los Angeles Owens River Aqueduct System, which was placed into service in 1913, to bring snowmelt from the east side of the Sierra Nevada (mountain range). As the need for water continued to grow, Mulholland and the LADWP initiated the water rights and preliminary planning for the Colorado River Aqueduct (CRA). The CRA was transferred for construction and operation to the Metropolitan Water District of Southern California (MWD), a state-created wholesale water agency, in 1928 and placed into service in 1941. MWD is also a contractor to receive water from the California State Water Project, which has brought water from Northern California since 1971. The Los Angeles Aqueduct has been expanded and modified several times since it was initially completed, including an extension of the aqueduct into the Mono Basin in 1940 and completion of the Second Los Angeles Aqueduct in 1970, but further descriptions of these and other modifications are beyond the scope of this report. The entire aqueduct system will be referred to herein as the Los Angeles Aqueduct (LAA) System. Today the citizens and industries of Los Angeles receive their water from the Los Angeles River ground water and other basins, LAA, MWD, and recycled water.

Figure 1. William Mulholland, first LADWP Chief Engineer and General Manager.

This report overviews the multihazard mitigation and improvements implemented in the Los Angeles Water System. A description of the Water System developments as related to multihazard mitigation is first presented, followed by descriptions of improvements and mitigations for different hazards effecting the Water System, identified by hazard type. Seismic and water quality concerns have dominated mitigation efforts, but in many cases were implemented with considerations of other hazards. Throughout its history the LADWP has been fortunate enough to have "champions" to support and promote mutihazard mitigation efforts.

WATER SYSTEM DEVELOPMENTS RELATED TO MULTIHAZARD IMPROVEMENTS

Development of the Los Angeles Water System has incorporated multihazard concerns from the initial inception of the LAA to present day modifications. The system layout was determined from normal operational and engineering requirements. Due to the need for daily water deliveries, designs naturally followed operational concerns. However, the LADWP has instituted for over 100 years a culture of developing a redundant system whenever and wherever possible to allow versatility under normal daily operational demands and emergency conditions. These redundancies have helped create a water system that is very versatile and resilient to effects resulting from manmade and natural hazards.

Figure 2. California map showing LADWP water supply sources and San Andreas Fault.

Supply Sources

The LADWP has developed many water supply sources to allow redundancy for normal and emergency operations. Figure 2 shows the supply sources from the LAA, CRA, California Aqueduct, and the Los Angeles River groundwater and other basins. In addition, recycled water is used to supplement potable water for irrigation and industrial purposes, but is not considered a primary supply in the event of a disaster. The numerous tanks and reservoirs contained in the water system store water from these various sources for distribution throughout the city. The numerous connections with the MWD throughout the City provide water from the CRA and California Aqueduct for distribution. In the event of a disaster, the water system draws upon the many available sources to provide a reliable water supply. Following a disaster, these supply redundancies allow water to be provided to unaffected areas of the city with little or no interruption, and also significantly aid in post-disaster recoveries to severely damaged portions of the water system. Within severely damaged portions of the system, the redundant supplies and in-city storage significantly reduce the outage time an average customer experiences following a disaster.

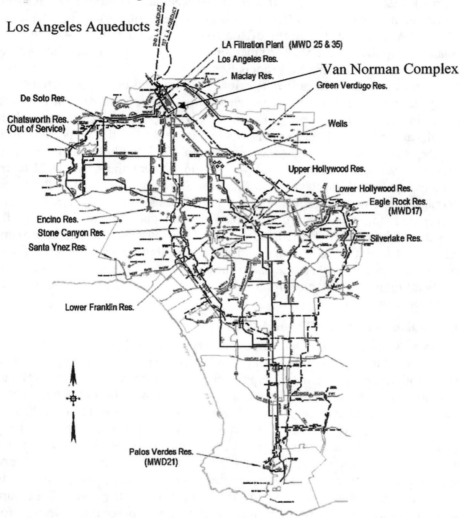

Figure 3. LADWP water supply and transmission system showing the first and second LAA entering the city from the north, major trunk lines (solid bold lines), MWD supply lines (dashed lines), reservoirs (labeled), major tanks (solid circles), other LADWP water facilities as labeled, and freeways (labeled solid lines), within the City boundaries.

Figure 3 shows the LADWP water supply and transmission system. Terminal storage for the LAA was developed at the City's northern limits, in an area now called the Van Norman Complex (VNC). The VNC, location shown in Figure 3, is an important site for collecting water supply and distributing it to the city because it serves as the LAA terminus and also receives the majority of MWD water. As the main water supply hub for the entire City, about 75 percent of the City's water passes through the VNC before it is distributed throughout the City. Once water leaves the VNC it is sent into portions of the distribution system and transmitted to reservoirs and tanks throughout the City.

Storage

The Lower and Upper San Fernando Dams, which retained the Lower and Upper Van Norman Reservoirs, were constructed in the VNC in 1913 and 1921, respectively, to provide terminal storage for the LAA. All other reservoirs used for normal operation along the LAA were located north of the San Andreas Fault. For seismic concerns Mulholland recognized the need for emergency storage south of the San Andreas Fault and as a result constructed the St. Francis Reservoir, which is described in more detail in the section on earthquake hazard mitigation. Mulholland also recognized the need for emergency storage within the City, which resulted in the construction of several large reservoirs including Hollywood, Franklin Canyon, Stone Canyon, Silver Lake, Chatsworth, and Encino Reservoirs shown in Figure 3, and the Upper and Lower Van Norman Reservoirs on the VNC (not shown). In 1924 Mulholland stated his desire to have reservoir storage holding a total 1-year supply near the City [1][2] to provide a buffer against potential shortages resulting from drought, earthquake, flood and debris flow damage to LAA, damage resulting from the Owens Valley "Water Wars", and to be better prepared for the continuing rapid growth that began following the LAA completion. At its peak in 1971, the Los Angeles Water System had approximately 24 large reservoirs to maximize storage and water distribution within the City for normal and emergency operations. At the same time, the City's water system contained eight smaller operating reservoirs, numerous water tanks, and six large water storage reservoirs along the Northern LAA.

Supply Transmission Lines/Trunk Lines.

Figure 3 shows the LADWP water supply system within the City limits. In order to provide a reliable water supply throughout the system, numerous large diameter [up to 120-inches (3048 mm) in diameter] supply trunk lines originate from the VNC and other supply sources and extend to various reservoirs and tanks throughout the City. Other trunk lines interconnect the main supply lines to allow for distribution of different water sources to many parts of the City. The trunk lines are also equipped with many valves that allow isolation of the water supply. The trunk line interconnectivity and isolation capability provides great versatility in the ability to supply water to many parts of the system from many sources. In many cases, damaged trunk lines can be isolated for repair without a significant water outage. At the same time, water can still be supplied to damaged portions of the distribution system from other sources in order to provide the needed water pressure for finding and repairing pipe breaks. The supply line redundancy is a significant component in creating a system resilient to severe hazardous effects, which was proven, for example, to be effective in the 1971 San Fernando and 1994 Northridge earthquakes.

Pressure Zones

The Los Angeles distribution system is broken into approximately 115 water pressure service zones. Water pressure in the different zones is controlled by the reservoir elevations, pressure head created by pumping, and pressure drop created by regulating stations. A regulator station causes a desired head loss to prevent high service pressures from entering lower ground elevation regions. The numerous pressure zones are the result of natural topographic changes throughout Los Angeles. Many pressure zones in Los Angeles can be supplied by multiple sources, multiple storage facilities, and multiple supply lines. Those that have a single source of supply are generally very small zones. The numerous pressure zones aid in the ability to isolate and provide redundancies to block areas of the distribution system that might be damaged in a disaster, such as an earthquake, which enhances the system's resiliency and reliability.

EARTHQUAKE

Los Angeles and the Owens Valley are located in highly seismically active areas. Numerous active and potentially active faults expose water system facilities to severe seismic hazards. Many earthquakes have damaged Los Angeles water facilities.

Water System Developments Related to Seismic Improvements

The Los Angeles Water System development has incorporated seismic concerns from the initial inception of the LAA to present day modifications. System components were designed with the common seismic considerations at the time of development, and many were developed with leading edge seismic technologies.

The LAA planning and design was underway when the 1906 San Francisco Earthquake and subsequent fire occurred. At that time Mulholland was also a consulting engineer for the Spring Valley Water Project, a system providing water to San Francisco which became the Hetch Hetchy aqueduct system. Mulholland studied the earthquake effects on the San Francisco Water System. He understood that the majority of property loss resulted from water system failure and resulting fire and also noted that all of the Spring Valley system earthen and concrete dams had remained in tact even though they were retaining full reservoirs and located close to the earthquake source. He also noted that some very important buried trunk lines at Millbrae-Belmont remained in tact, which helped ensure the City of San Francisco was not entirely deprived of water. These experiences along with studies of the 1872 Owens Valley and 1857 Fort Tejon Earthquakes influenced seismic development of the LAA, water storage, and distribution systems [1]. The following two subsections summarize some seismic considerations incorporated in the early developments and knowledge gained during implementation that helped further the general understanding of earthquakes and water systems.

Supply

In consideration of earthquake effects on the LAA, Mulholland noted that most of the aqueduct would be buried, "little need be feared as to any irreparable destruction due to seismic shock," and although admitting damage could occur, he contended that it would be confined to local breakage here and there along the aqueduct line and "would not be beyond the possibility of repair in reasonable time and at a moderate cost". This concept of post-earthquake repair as a mitigation strategy has worked for many damaging earthquakes beginning soon after the LAA was placed in service; on July 11, 1917 an earthquake damaged the LAA in the Alabama Hills and was apparently was related to tunnel damage near Little Lake [1].

The LAA passes through the San Andreas Fault zone in the Elizabeth Lake Tunnel. In Mulholland's lecture to mining engineers in 1918, he described his observations of the San Andreas Fault during the Elizabeth Lake Tunnel excavation, which included a scholarly description of the rock formations encountered during the tunnel excavation [1]. According to J. David Rogers, a geologic engineer who studied Mulholland's engineering career [2], "it is the first such account ever given on an active fault underground at considerable depth 400 feet (122 m) beneath the ground surface"[3].

Storage

As shown in Figure 2, the San Andreas Fault generally runs from southeast to northwest nearly the entire length of the State of California, and the LAA intersects the fault about 25 miles (40 km) north of the City. William Mulholland recognized that movement on the San Andreas Fault could sever the LAA at the Elizabeth Lake Tunnel. As a result, in addition to the normal operational storage, Mulholland identified emergency storage requirements south of the San Andreas Fault, which was met in 1926 by construction of the St. Francis Reservoir in San Francisquito Canyon, shown in Figure 4, and its associated hydroelectric power plants [4]. Several reservoirs were also constructed within the City. Unfortunately, as shown in Figure 5, the St. Francis Dam failed catastrophically in March of 1928. After the dam failure, the need for LAA emergency storage south of the San Andreas Fault still existed and Bouquet Canyon Reservoir, shown in Figure 6, was constructed to fulfill the storage needs [5].

Figure 4. St. Francis Dam and Reservoir placed in service in 1926 to provide water storage south of the San Andreas Fault.

Figure 5. 1928 failure of St. Francis Dam.

Figure 6. Bouquet Canyon Reservoir constructed to provide replacement storage south of the San Andreas Fault after the St. Francis Dam failure.

Seismic Improvements Prior to the 1971 San Fernando Earthquake

The most significant Water System seismic improvements, outside of those described as part of the system developments, were initiated following the 1971 San Fernando Earthquake as a result of lessons learned from earthquake impacts on modern urban infrastructures. However, there are several important examples of significant seismic related activities and improvements implemented prior to the 1971 earthquake.

Building Code

The first edition of the Uniform Building Code was published in 1927 and contained an optional appendix prescribing a lateral force coefficient of 0.1g. Following the 1933 Long Beach, California earthquake, the City of Los Angeles, County of Los Angeles, City of Long Beach and other municipalities, primarily in Southern California, adopted seismic provisions that commonly used a lateral force coefficient of 10% of gravity, following the precedent established in Japan. Since the LADWP construction was in the City of Los Angeles, it was required to obtain City building permits that complied with similar seismic codes.

Since 1933, all Water System structures have been designed to meet the seismic codes in force at the time with the addition of an importance factor when appropriate, which increases the minimum code seismic coefficient. The importance factor is applied by recognizing the need for the facility to be operable after a seismic event. In more recent times, many critical facilities have been designed using site specific seismic criteria, exceeding minimum code requirements.

Strong Motion Instrumentation

In the 1950's early versions of strong motion instruments, called Wilmont Seismoscopes shown in Figure 7, were installed at a number of LADWP dams. The seismoscopes were usually placed in pairs with one instrument on the dam crest near the center and the other on the dam abutment. These instruments record motion of a pendulum with a needle that scratches a blackened hourglass located at the top of the pendulum. Several seismoscopes are maintained and presently remain in service.

Figure 7. Wilmont Seismoscope strong motion instrument.

Strong motion instrumentation and data collection was implemented in 1965 to evaluate the performance of buildings and structures in the immediate aftermath of an earthquake. Code-mandated instrumentation about that time was required for most buildings over six stories high in certain California cities like Los Angeles, which adopted the instrumentation provisions of the Uniform Building Code [6]. The LADWP General Office Building (John Ferraro Building, JFB) was under construction at that time. Strong motion accelerographs (SMA) were installed in the JFB basement, 7th floor, and top floor. These were the first instruments to be installed in the City and the first to be manufactured by Kinemetrics. The first SMA-1 instrument, shown in Figure 8, installed by Kinemetrics was in the JFB and is now preserved by Kinemetrics and displayed in their Pasadena, California front office.

Figure 8. The first Kinemetrics SMA-1 strong motion instrument (Kinemetrics).

Many important LADWP sites have been instrumented with strong and weak motion recorders for seismological and engineering purposes. These instruments have been installed by the LADWP or by other organizations with cooperation from the LADWP, including other government organizations such as the California Geological Survey, United States Geological Survey, universities such as the California Institute of Technology, University of Southern California, and earthquake research centers such as the Southern California Earthquake Center (SCEC). Instruments such as the SMA-1 in Figure 8 are installed at many LADWP dam sites. Information obtained from installations on LADWP sites has made very significant contributions in seismological research and earthquake engineering developments; most notable are the LADWP recordings made during the 1994 Northridge Earthquake [7].

Soil Mechanics

Ralph R. Proctor, shown in Figure 9, was a LADWP Water System Civil Engineer and innovator of modern geotechnical engineering and soil mechanics. The methods for soil compaction and testing developed by Proctor have proved to be very useful in improving the seismic performance of geotechnical structures, such as embankment dams and fills, and building foundations throughout the world. Many LADWP dams constructed before 1928 used hydraulic fill methods for placing soil, which left relatively loose and weak soil deposits making up the embankment dam. A few dams built in the 1920's used newly emerging mechanical compaction methods, which provided some strength improvement over hydraulic fill, but both hydraulic fill and mechanical fill placement methods were used through the 1920's.

Figure 9. Ralph R. Proctor, inventor of modern soil compaction and testing methods.

The LADWP recognized the need for improved dam construction techniques. With the advent of improved mechanical equipment, Proctor developed specifications to construct heavy sheepsfoot rollers and methods for placing and compacting soil for the primary purpose of constructing embankment dams [8], but the methods are useful for any compacted earthen structure. He also developed testing methods and equipment to monitor construction of dams and to measure the in-place density of soil in the field and the shear strength of soil in the laboratory. Additionally, Proctor developed a specialized tool called the Plasticity Needle, now commonly called the Proctor Needle that was used for measuring penetration resistance to indirectly measure field density, shear strength, and moisture content. The Plasticity Needle penetration resistance was calibrated in the laboratory using soil test specimens compacted at different moisture contents to allow rapid identification of soil density and moisture content in the field. Figure 10 shows a partial set of needles with the penetrometer and the laboratory compaction mold and rammer developed by Proctor. The use of needles for aiding in compaction testing has been standardized by the American Society for Testing and Materials (ASTM) as standard ASTM D1558.

Figure 11 shows the initial implementation of Proctor's methods during the construction of Bouquet Canyon Reservoir Dams No. 1 and 2 (1932-1934), which is the first large scale use of modern soil compaction with sheepsfoot rollers and testing methods; these two dams are the first ever to be constructed using rigorous quality control techniques. As shown in Figure 11, for these dams the sheepsfoot rollers, which were designed and constructed by LADWP forces, followed a single pattern until the soil had been rolled a minimum of 16 times. Figure 11a shows the compaction process and equipment used to dump, spread, and compact the soil. Figure 11b shows an extra heavy LADWP-designed and fabricated sheepsfoot tandem roller. Bouquet Canyon Reservoir was constructed primarily for seismic improvements to the water system in order to provide emergency water storage along the LAA south of the San Andreas Fault to protect the City's water supply in the event of a major earthquake. Modifications of the procedures for testing relative soil compaction developed by Proctor have been standardized in ASTM D698 and ASTM D1557, commonly referred to as the Standard Proctor and Modified Proctor testing procedures, respectively. The Proctor tests, and the methods developed at the LADWP for placing and compacting soils, are today international standards for improving soil performance for every day working loads and seismic concerns, and may possibly be the single most significant factor for improving the seismic resistance of engineered structures worldwide.

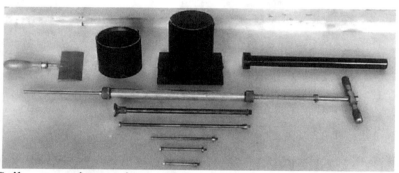

Figure 10. Soil compaction testing equipment: compaction mold (top center), rammer (top right), cutting tool (top left) and Proctor Penetrometer and Needles (bottom).

a.   b.

Figure 11. Soil compaction for Bouquet Canyon Reservoir Dam Number 1.

It should be emphasized that the soil compaction methods developed by Proctor were primarily for improving dam safety and performance in the aftermath of the St. Francis Dam failure. The seismic consideration is only one aspect of dam performance. Proctor also developed and implemented improved seepage control methods for dams. These items are truly multihazard improvements and are identified in this section for the primary reason that they were initially developed for Bouquet Canyon Reservoir to provide emergency water storage in the event of an earthquake, and because soil compaction and seepage control have since become prominent geotechnical improvement methods for earthquake resistance.

Embankment Dams

Since the mid-1930's, all LADWP dams were constructed of well compacted earth fill with special seepage control measures; most of which were built under Proctor's supervision. Stone Canyon and Encino Dams were originally constructed in the 1920's using combinations of hydraulic and newly emerging mechanical rolled fill methods, respectively. In the 1950's and 1960's, the capacities of these two reservoirs were increased to provide additional in-city water storage. As part of the reservoir improvement programs, the new embankments were designed to have stable slopes using a 0.1g lateral seismic coefficient and a significant volume of the existing embankment dams and underlying alluvium were excavated prior to placing well-controlled compacted earth fill. The purpose for removing the existing dam and alluvium materials were to help ensure that the new embankments were able to resist strong seismic shaking. For both Encino and Stone Canyon Dams, portions of the existing loose, relatively weak dam fill and alluvium were left under the downstream slopes of the re-constructed dams. All of the dams constructed and significantly modified since 1930 have proven adequate to resist seismic forces under actual shaking and using modern state-of-the-art analyses. Therefore, the advancing dam design and construction methods significantly improved seismic stability. LADWP earth fill dams constructed between 1930 and 1971 were conservatively designed and constructed with seismic stability in mind. LADWP records indicate that dams designed at least since the early 1950's, such as Stone Canyon Dam, were evaluated for seismic slope stability using a minimum of 0.1g horizontal coefficient. No records have been identified to show that earlier dams, such as the Bouquet Reservoir dams in the early 1930's, were designed with seismic stability analyses as we use today; these seismic improvements are considered to be more intuitive than analytical.

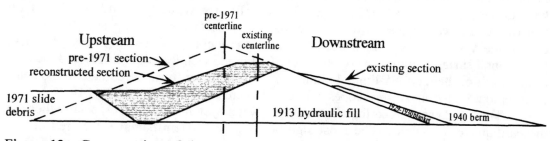

Figure 12. Cross-section of the Lower San Fernando Dam showing the original 1913 hydraulic fill, 1940 downstream berm, cross-section geometry in 1971 prior to upstream slope failure, and reconstructed shape following 1971 earthquake.

Figure 12 shows a cross-section of the Lower San Fernando Dam (LSFD). Stability concerns for the LSFD in the 1930's led to the placement of a compacted fill buttress on the downstream slope. The LSFD retained the Lower Van Norman Reservoir and was primarily constructed in 1913 using standard hydraulic fill methods. The downstream buttress fill was placed in 1940. The buttress fill material was obtained from a large storm water improvement project undertaken on the VNC to bypass storm water around the west side of the Upper and Lower Van Norman Reservoirs [9].

The LSFD provides the first fully documented case in which a LADWP dam was analyzed for seismic slope stability [10]. The LSFD was evaluated in the 1960's by LADWP engineers, under the review of an eminent board of consultants consisting chair Paul Bauman, Assistant Chief Engineer Los Angeles County Flood Control District, Charles F. Richter, California Institute of Technology, C. Martin Duke, University of California at Los Angeles, and Vladimer Pentigoff, formerly of the U.S Army Corps of Engineers. The evaluation used a pseudo-static lateral force coefficient, similar to common simplified slope stability methods used today. However, liquefaction was not considered to be a concern for embankment dams, mainly because little was understood about liquefaction at that time [11]; therefore, liquefaction was not considered as part of the seismic stability analysis. Nevertheless, the stability evaluation indicated the LSFD may have slope stability problems when shaken by a significant earthquake. As a result of this evaluation, a large seismic improvement project was implemented that consisted of constructing the Lower Van Norman Bypass Pipeline and Reservoir (Bypass Pipeline and Bypass Reservoir). Once the Bypass Pipeline and Reservoir were constructed, the LSFD was planned to be removed and replaced with a new compacted earth fill dam. The 60-inch (1524 mm) diameter Bypass Pipeline was constructed in 1968 and the 240 acre-foot (296,036 m^3) Bypass Reservoir was completed and placed in service in November 1970. The final phase of the seismic improvement project, reconstruction of the LSFD, was never completed as a result of a large liquefaction-induced slide on the LSFD upstream slope during the February 9, 1971 San Fernando Earthquake, only two months after placing the Bypass Reservoir in service. The new Bypass Pipeline and Dam, shown in Figure 14, were not damaged by the San Fernando Earthquake.

1971 San Fernando Earthquake

The February 9, 1971 San Fernando Earthquake occurred on the San Fernando Fault with a moment magnitude M_w 6.7 and epicentral distance approximately 7 miles (11 km) north-east of the VNC [12]. The VNC overlies the westerly boundary of the fault that ruptured toward the southwest. The main fault surface ruptures primarily occurred east of the VNC, with small surface ruptures in the vicinity of the LSFD east abutment [13]. Small surface ruptures were also identified in the vicinity of the LSFD and the area to the north of the dam. A limited number of seismic recordings were made in the near-fault region, including seismoscopes on the LSFD crest and east abutment [14].

The San Fernando earthquake damaged water system facilities, especially those in the earthquake near-field. The most significant water system damage occurred on the VNC [15] with liquefaction induced slides resulting on the Upper San Fernando Dam (USFD) and LSFD. Figure 13 shows damage to the LSFD, after the reservoir was partially drained; the upstream slope slid over 200-feet (61 m) into the reservoir and the dam lost 30-feet (9.1 m) of freeboard, leaving only 5-feet (1.5 m) of freeboard above the reservoir water surface after the earthquake. Draining of the lower reservoir began soon after the earthquake and the upper reservoir was permanently lowered. Loss of the valuable VNC storage in the Upper and Lower Van Norman Reservoirs revealed the importance and timeliness of the Bypass Pipeline and Reservoir seismic improvements; without these two facilities the water system would have had extreme difficulty in supplying water to the City.

Figure 13. Lower San Fernando Dam, looking to the west from the east abutment area, showing upstream slope slide into the reservoir.

Additionally, breaks occurred on many large diameter trunk lines and several thousand distribution pipes and service connections throughout the system, primarily in the San Fernando Valley area. Damage also occurred to LAA pipes and channels, water storage tanks, older pumping and chlorination station buildings, unreinforced masonry buildings, water operating district yard buildings, and embankment dams in addition to the USFD and LSFD described above.

Post 1971 San Fernando Earthquake Seismic Improvements

Many lessons were learned from water system seismic damage in 1971. As a result, numerous seismic improvements were implemented in the years following the earthquake. The LADWP also significantly aided in the data collection and documentation of water system damage [16], allowing for great advances in lifeline earthquake engineering research and development [17] that made important contributions to LADWP and other organization system improvements worldwide. The seismic lessons, implementations, and mitigations are too numerous to describe in detail in this report. Only some highlights will be described herein.

Damage sustained to the USFD and LSFD resulted in an extensive research program funded by the LADWP and California Department of Water Resources Division of Safety of Dams (DSOD) that was undertaken by the University of California Berkeley under the direction of Dr. Harry Seed, Professor of Civil Engineering. The study results [11] identified liquefaction and slope stability as the primary cause of damage to the dams and made several recommendations for improved seismic stability evaluations of dams including advanced numerical computer modeling and dynamic laboratory testing.

As a consequence, the DSOD required all jurisdictional dams in the State of California to have seismic stability evaluations. The LADWP implemented recommendations from the study to the greatest extent possible and helped lead the geotechnical industry in dam stability evaluations throughout the 1970's. The LADWP obtained support from Dr. Seed and other prominent earthquake researchers, expert consultants, and review boards before developing final conclusions and implementing recommendations from the relatively advanced testing and analyses. Custom cyclic triaxial chambers were designed by LADWP engineers and fabricated in the LADWP machine shop to be used with an MTS Universal Electro-Hydraulic Testing Machine to simulate earthquake loading. State of the art finite element programs were used by LADWP engineers and consultants to analyze all DSOD jurisdictional dams. Stability analysis results showed that many existing dams were seismically stable and identified several dams that were in need of improvement or replacement. All dam stability analyses were performed with site-specific seismic design criteria for Maximum Credible Earthquakes (MCE).

Figure 14. Van Norman Complex improvements before and after the 1971 San Fernando Earthquake; Photograph identifies the Lower Van Norman Bypass Reservoir, Los Angeles Reservoir, and reconstructed Lower San Fernando Dam.

An extensive dam improvement program was undertaken and seismic deficiencies for dams on at least 12 reservoirs were mitigated by removing and reconstructing the dams, removing the dams from service, or restricting the reservoir high water level. In 1979 the Los Angeles Reservoir, shown in Figure 14, was placed into service on the VNC to replace the original Upper and Lower Van Norman Reservoirs. In the 1980's the LADWP worked with consultants from Dames and Moore to complete the first practical non-linear effective-stress deformation analyses of the Pleasant Valley and South Haiwee Dams [18][19][20], which are both located in the Owens Valley.

The 1971 San Fernando earthquake also caused damage to some of the older pumping stations, chlorinating stations and water operating district yard buildings. Following the 1971 earthquake, a seismic water system vulnerability assessment (SVA) was made under the direction of the lead author for all pumping stations, chlorination stations, maintenance and construction yards, tanks and sumps, reservoirs, and well installations in the City [21]. Figure 15 shows a SVA example for pumping stations excerpted from [21]. Surveyors and construction inspectors made the vulnerability assessments using visual inspections, after being trained to look for anchorage of equipment, type of building construction, roof to wall connections, hazardous materials storage, electric substation facilities, flexible pipe-to-structure connections, etc.

Questionable building construction and other seismic deficiencies were referred to the appropriate engineers for evaluation. The facilities were graded as low, medium, or high for importance to system and cost. A similar SVA report was prepared in the Owens Valley and the Mono Basin. The SVA reports were used for a multi-year budgeting program to implement the facility seismic upgrades. Some of these facilities were seismically retrofitted and others were completely reconstructed. All pumping station, chlorination station, and district yard buildings were evaluated and improved as necessary.

PUMPING STATIONS

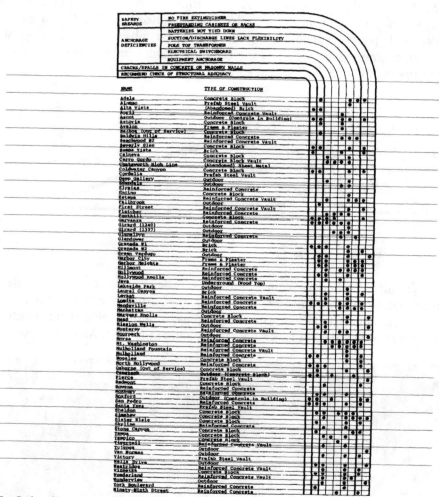

Figure 15. Seismic vulnerability assessment for pumping stations performed after the 1971 San Fernando Earthquake [21].

Following the 1971 earthquake, two large diameter pipes were retrofitted to help mitigate future earthquake damage. These mitigations were undertaken on the Second Los Angeles Aqueduct (SLAA) at Terminal Hill [22] and the Granada Trunk Line (GTL) on the VNC [23][24] as a direct result of damage inflicted by the 1971 earthquake. The

SLAA was damaged from a unique combination of large ridge shattering and slope deformations. Special pile supported piers were installed to support the pipe traversing the slope, flexible couplings were installed at the top of the hill, and rock anchors, as shown in Figure 16, were installed to restrain future ridge movements [25]. The GTL suffered damage to many pipe joints and mechanical couplings as a result of large liquefaction induced ground movements [23][24]. Figure 17 shows examples of wrinkled bell and spigot welded steel joints that occurred on the GTL, SLAA, and other pipes. Mitigation of the GTL to resist future ground deformations included installation of 12 special unrestrained long barrel mechanical couplings, shown in Figure 18, designed and constructed by LADWP personnel. A few trunk lines constructed after the 1971 earthquake to supply water from the VNC were routed specifically to avoid known liquefaction and lateral spreading areas that were observed following the 1971 earthquake; an example is provided by the Foothill Trunk Line construction in the 1980's around the "Juvenile Hall Slide" lateral spread.

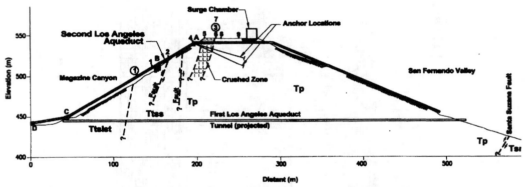

Figure 16. Terminal Hill cross-section showing crushed/shattered rock zone, damage and displacement locations from the 1971 San Fernando Earthquake (circled numbers), post-1971 tie-back anchor installation, and 1994 damage and displacement locations (non-circled numbers) on the Second Los Angeles Aqueduct pipeline. Modified from [25].

a.   b.

Figure 17. 1971 compression wrinkling of welded bell and spigot joint; a) pipe joint in place, b) cut out of wrinkled pipe joint.

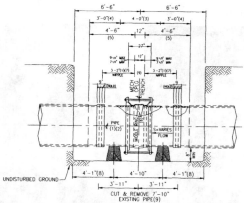

Figure 18. LADWP long barrel mechanical coupling design for the Granada Trunk Line (GTL) to accommodate displacements from permanent ground movements.

1994 Northridge Earthquake

The January 17, 1994 Northridge Earthquake occurred on an unmapped blind thrust fault with a M_w 6.7 and an epicenter located in the Northridge/Reseda area of the San Fernando Valley [12]. The ruptured thrust fault was below much of the San Fernando Valley, which subjected many critical water system facilities to strong near-source pulses. Many seismic recordings were made in the near-fault region, including several important recordings at LADWP facilities on the VNC [7], which included some of the largest ground motions ever recorded.

The Northridge earthquake caused considerable damage to water system facilities, especially those in the earthquake near-field, but not nearly as significant as the damage caused by the 1971 earthquake. Many breaks occurred on large diameter trunk lines and several thousand distribution pipe and service connection leaks resulted, primarily in the San Fernando Valley [26]. Damage also occurred to LAA pipes and channels, water storage tanks, and some buildings [27]. Figure 19 shows water flowing from the GTL in Balboa Boulevard. The leak resulted from ground deformations [28] causing damage to the pipes as shown in Figure 20. A natural gas line was also damaged, which ignited after a car stalled in the water.

Figure 19. Water flowing into Balboa Boulevard from damaged Granada Trunk Line (GTL). A fire ignited from a natural gas pipe leak (photo by Doug Honegger).

a.

b.

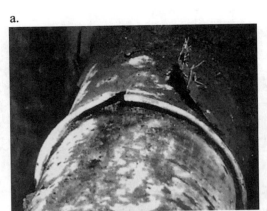

Figure 20. 1994 damage to Granada Trunk Line (GTL) in Balboa Boulevard resulting from permanent ground deformations: a. compression wrinkling on south end of damage zone, b. tension separation on north end of damage zone.

As was the case during the 1971 San Fernando Earthquake, the greatest water system damage in 1994 occurred in the northern San Fernando Valley, in and around the VNC [29]. The most significant problems were to water supply pipelines and channels. Restoration of these damages was critical for restoring water supply before the in-city storage was depleted. There was some damage to embankment dams, but not significant enough to affect operations, except for a small dike on the San Fernando Power Plant Tailrace, shown in Figure 21, which failed and affected supply from the LAA [30]. Several older tanks were damaged in the Santa Susana and Santa Monica Mountains [31].

Figure 21. Power plant tailrace dike failure following the 1994 Northridge Earthquake.

Nearly all of the improvements made to building structures, tanks, and dams following the 1971 earthquake, and all new facilities constructed after 1971 performed well in the 1994 earthquake, proving that the lessons learned and mitigation strategies implemented were very beneficial. However, there were some additional lessons learned from the 1994 earthquake, including the identification of necessary improvements to buried pipes and the effects of near-source ground motions on above ground and buried structures and geotechnical structures. All of the damage, lessons learned, and seismic mitigations resulting from the 1994 earthquake are too numerous to report herein. Only a few highlights are described below.

Post Northridge Earthquake Seismic Improvements

Seismic improvements following the 1994 earthquake consisted of (1) implementing knowledge and design strategies that had previously been learned from the 1971 earthquake, but not fully carried out on all facilities due to priority and funding conflicts with water quality and other issues, and (2) identification of new earthquake related issues requiring further research.

An example of implementing previous lessons is provided in tank performances. Many damages to older tanks, which included breakage of rigid inlet-outlet lines, roof damage, and movement of unanchored tanks, were of no great surprise after evaluating the post-earthquake damage. Much of this damage type was identified in the 1974 SVA report [21] and recommended to be reviewed further. In addition, knowledge gained and utilized on tank performance and design were proven effective by the fact that all newer tanks performed well [31]. As a result of damage to older tanks, an improvement program was undertaken after the 1994 earthquake to add flexible connections to tank inlet-outlet lines where needed that allow extension, compression, and rotation of the connections. Figure 22 shows one type of flexible connections used on tanks.

a.   b.

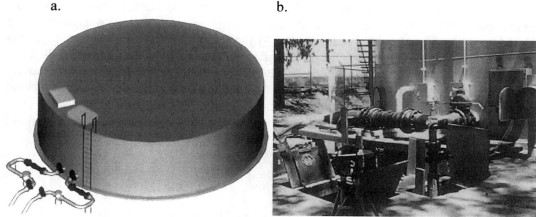

Figure 22. Example of tank inlet-outlet line flexible connections: a) three-dimensional rendering of tank and flexible inlet-outlet connections; b) installation of flexible connections at Harbor Heights Tank No. 2.

Examples of mitigations and improvements resulting from new knowledge learned from the 1994 earthquake include:

1) Figure 23 shows failure of the 96-inch (2438 mm) diameter corrugated metal pipe Lower San Fernando Drain Line No. 1 after it sustained a complete lateral collapse. This pipe was replaced with a reinforced concrete pipe surrounded by a heavily reinforced concrete encasement. The LADWP and the University of Southern California (USC) performed research on this unique pipe failure and determined that the combination of dynamic pore pressure build up in the surrounding bedding soil and large near source shear strain pulses led to this damage [32][33][34].

a.   b.

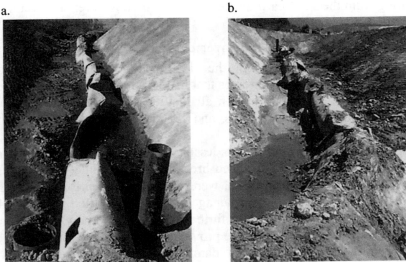

Figure 23. Failure of Lower San Fernando Drain Line No. 1 corrugated metal pipe. a. view looking south, b. view looking north.

2) The SLAA at Terminal Hill and GTL on the VNC both sustained damages during the 1994 earthquake, even though significant mitigation measures were undertaken following the 1971 earthquake. Investigations were carried out to better identify causes of damages to these facilities [23][24][25]. Additional mitigation strategies were evaluated and in each case, relocation was determined the best alternative. A seismic mitigation project relocating the GTL out of the liquefaction induced ground deformation zone has been completed. Relocation of the SLAA through a 600-foot (183 m) tunnel and a 300-foot (91 m) vertical shaft at Terminal Hill is currently in design and planned to be completed by 2006.

a.   b.

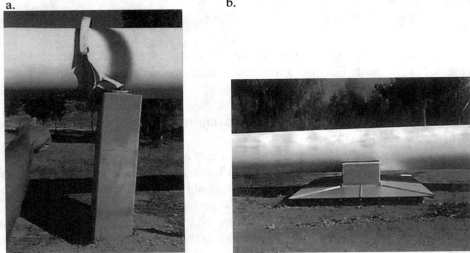

Figure 24. Van Norman Pumping Station Discharge Line: a. Earthquake damage to pier foundations and ring girders, and b. Completed base isolation pad and protective roof.

3) Figure 24a shows the 54-inch (1372 mm) diameter Van Norman Pumping Station Discharge Line, an above ground welded steel pipeline supported by ring girders on concrete piers and steel H-piles. The pier supports for this pipe were originally damaged in 1971, but did not require repairs; instead they were left in their post-San Fernando Earthquake tilted condition. The pipe did not leak and damage was limited to leaning of piers, which continued performing their intended function, and considered to be a minor structural problem. In 1994, the piers were further damaged; causing significant leaning and failing several ring girders, but the pipe did not leak. Follow-up investigations identified liquefaction and lateral ground movement as the mechanism damaging the piers and underlying H-pile foundations. Figure 24b shows the pipe after completion of a seismic repair and mitigation project. Figure 25 shows details of the seismic mitigation designed and constructed by LADWP engineers and construction crews led by the second author, which consisted of lowering the pipe near the ground surface and placing the supports on low-friction base isolators made of Teflon and stainless steel. This is the first known use of base isolation to mitigate liquefaction-induced ground deformation effects to a water pipeline.

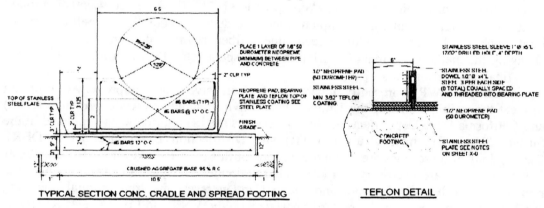

Figure 25. Details of LADWP designed Teflon bearing surface base isolator for the Van Norman Pumping Station Discharge Line.

4) The High Speed and Bypass Channels are important LAA water conduits, which also transport a significant amount of California Aqueduct water from MWD connections, on the VNC that were damaged in 1971 and 1994 from permanent ground deformations in areas where the channels pass through weak, saturated natural soil deposits [29][35]. Case studies from multiple earthquakes identified the differing effects that transient and permanent ground movements had on channel damage and that there is possibly a local permanent deformation regime causing different levels of channel damage within the larger lateral ground deformations that extend well beyond the channels [36][37]. The case study results are currently being evaluated in greater detail to determine the feasibility of performing soil improvement techniques to cost effectively mitigate channel damage for future earthquakes.

As a direct result of the Northridge Earthquake, the LADWP has significantly aided the research and development of new technologies and advancement in earthquake engineering knowledge through cooperative research with the University of Southern

California (USC), Southern California Earthquake Center (SCEC), Pacific Earthquake Engineering Research Center (PEER), Multidisciplinary Center for Earthquake Engineering Research (MCEER), the United States Geological Survey, and others. The numerous cooperative research undertakings have significantly helped develop knowledge and understanding in near-source ground motions [38][39][40], understanding of underground pipe behaviors [33], permanent ground deformation measurements [41], development of liquefaction and lateral spreading databases and improved models [42], development of GIS-based earthquake evaluations and modeling [41][42][43][44], water system component fragility curves [45][46] and other projects that have improved seismology and earthquake engineering practices.

Present and Future Perspectives

The LADWP continues to provide support and leadership in advancing earthquake engineering practices. For example, LADWP provided numerous pipe samples and support to MCEER for testing and modeling of welded bell and spigot steel pipe joints to better understand the mechanism leading to buckling, as shown in Figures 16 and 20a, of these joints under axial compression [47]. The research was then applied to develop strategies that can be implemented to reduce the potential for bell and spigot joint buckling by using welding techniques, fiber reinforced composite wrapping, changing the methods for manufacturing the joint shape, etc. Research for this project is in its final completion stages [48].

The LADWP is presently undertaking an extensive capital improvement program to meet the requirements of the United States Environmental Protection Agency (EPA) and California State Department of Health Services (DHS) requirements stipulated in the Surface Water Treatment Rule (SWTR) and Disinfection Byproducts Rule (DBR). Descriptions of these rules are beyond the scope of this report; however significant Water System changes are necessary to meet their requirements. System changes include the removal of Hollywood, Encino, and Stone Canyon reservoirs from normal operating service, which places a much greater importance on the VNC and Los Angeles Reservoir for water supply throughout the City on a daily basis. These three reservoirs are planned to retain storage for emergency supply purposes. Hollywood and Encino Reservoirs have already been removed from service and Stone Canyon Reservoir is scheduled to be removed in 2004. Silver Lake Reservoir is planned to be similarly removed from normal operating service in the near future and replaced with alternate covered storage, of reduced capacity. The Los Angeles Reservoir is planned to be divided in half with a new embankment dam to allow for greater system flexibility and each side will be covered with a floating cover. In addition, many miles of new large diameter trunk lines are being installed to allow greater system flexibility for the water quality projects and also as a part of a trunk line replacement program. The replacement program was initiated following a study of older trunk lines which identified several that were in need of repair and replacement.

The system changes necessary for water quality improvements leave ambiguity concerning how the system may perform in future earthquake scenarios similar to what the City experienced in 1971 and 1994. Although in-city storage capacity is being significantly reduced the increased number of trunk lines being constructed will provide greater redundancy and flexibility. As a result, the Board of Water and Power

Commissioners (Board) requested the Water System to be evaluated for its ability to withstand natural disasters.

In response to the Board's request, a dam seismic stability evaluation program is underway to evaluate the performance of 17 of the LADWP's large dams that are located south of the San Andreas Fault. The initial phase included a screening evaluation of the dams to determine their seismic stability using current knowledge of Southern California seismicity and state-of-the-practice simplified analysis methods [49]. All of these dams have previously been evaluated and determined to be seismically stable. This new program is to re-evaluate the dams based on the most recent knowledge. The screening evaluation identified 8 dams that needed additional information from field and laboratory tests and more detailed evaluations, which are currently underway and planned to be completed in a few years. During the more rigorous analysis of Stone Canyon Dam, LADWP engineers worked closely with consultants from URS Corporation to characterize the extent of the alluvium left in place under the downstream shell of the dam and to complete the first practical 3-dimentional non-linear effective-stress deformation analysis of an earth dam [50][51].

In 2002, the LADWP initiated a cooperative program with MCEER to perform an extensive research project for developing state-of-the-art system analysis modeling techniques that can account for non-linear hydraulic flow conditions to model pipe breaks in the system. The programs developed will be calibrated with known performances and actual recorded data from the Northridge Earthquake. The models are planned to be utilized for different seismic hazards throughout the Los Angeles area to estimate system performance under other earthquake scenarios, which can then be used to help identify potential system weaknesses and also help in emergency response and recovery.

The LADWP is also initiating a seismic hazard evaluation program for the LAA. Except for dams, there have been no significant seismic mitigation strategies implemented along the LAA; as previously described, the seismic strategy for the LAA implemented by Mulholland was to perform repairs following an earthquake. However, the two aqueducts do provide redundancy for emergency operations. Damage from the 1971 and 1994 earthquakes required time consuming and extensive repairs that caused concern and in some cases a desire for a more prompt return of the LAA water supply under emergency conditions. As a result some previously described seismic improvements are underway within City limits. These experiences are cause for concern that large earthquakes on faults traversing the LAA may develop significantly more damage and impede LAA flow sufficiently to impact the overall available water resources to the City.

The Sepulveda Trunk Line (STL) is a new 84 to 96-inch (2134 to 2438 mm) diameter trunk line that was put in service in July 2003, to supply water from the VNC to many other parts of the City. As part of the system changes, the STL will become one of the most critical and important supply pipe lines in the City. The STL crosses the San Fernando Fault zone, a westerly extension of the fault that ruptured in 1971. As a result of this active fault hazard, the STL was aligned to cross through the fault zone as rapidly as possible and constructed with welded butt joints within the fault zone. Isolation valves were installed outside of the fault zone and additional interconnections were made to other nearby trunk lines. In addition, the STL was originally intended, in part, to replace the old 1913 vintage riveted steel City Trunk Line as part of the replacement program.

After evaluating the seismic threat to the critical STL pipeline, plans for the City Trunk Line were altered to keep it in service and have it slip-lined with a high density polyethylene (HDPE) pipe, of smaller diameter than the original riveted steel carrier pipe. This will provide additional redundancy to the STL as they both pass through the same fault zone. The City Trunk Line is supported on piers through a portion of the fault zone.

The Pacific Pipeline is a privately owned pipeline that transports crude oil from Central to Southern California to several refineries. The pipeline crosses over the LAA channel within the active Santa Susana fault zone. As a part of approval to construct the new oil pipe over the LAA, several improvements were implemented to help resist seismic forces, including constructing an embankment berm around and covering the LAA channel in the pipe vicinity, routing the oil pipe above ground as it passes over the channel and placing it within a larger diameter pipe set within an even larger reinforced concrete box structure to aid in resisting strain from potential fault movement and help contain and control any potential leakage. Consultants from Woodward-Clyde and an expert engineering panel experienced with fault offset concerns for critical pipelines provided advice and direction in implementing the Pacific Pipeline improvements.

As part of the water quality improvements and the numerous large diameter pipelines and appurtenant underground structures being constructed, the LADWP recently developed and implemented an improved method for evaluating lateral seismic stresses on rigid underground structures such as vaults and box conduits [52]. The method is much simpler to implement, provides more realistic pressures than previous models, and is easy to implement into common underground structural design procedures.

Earthquake Prediction

Earthquake prediction is a very young and evolving science. There are several meaningless earthquake predictions per year that statistically will occur regardless of the predicted condition and do not affect the Water System beyond standard earthquake preparations. However, occasionally there are predictions with a good scientific basis [53] identifying a time of increased probability that an earthquake may occur. For these predictions the Los Angeles Water System may take certain publicly responsible pre-operational precautions. The precautions may vary depending on the earthquake prediction zone, but include updating emergency response procedures, lowering certain reservoirs for dam safety concerns, increasing in-city water storage in certain areas, some maintenance procedures, and encouraging employees to update their personal earthquake preparedness. Earthquake prediction has not advanced to a level to warrant prioritization of capital expenditures for seismic mitigation based on the prediction.

PIPELINE DETERIORATION AND REHABILITATION

Old deteriorated water pipes expose the Los Angeles Water System to hazards resulting from reduced hydraulic capacity, water quality, and pipe leaks and ruptures. A major portion of the Los Angeles water distribution pipeline system was installed before the use of cement mortar lining to prevent interior corrosion. Interior corrosion causes reduction in the hydraulic capacity of the pipeline, lead to failure, and can cause water quality problems. Since the 1940's pipelines were purchased with an interior cement mortar lining. Many old large diameter trunk lines in the City, some constructed nearly 100 years ago, are deteriorating and occasionally rupture, causing significant operational problems and property damage. A program has been undertaken to rehabilitate or replace the old trunk lines. Prioritization of at-risk trunk lines was established after performing a series of in depth studies of riveted and non-riveted steel pipes as a part of the trunk line condition assessment program [54][55]. Results of these studies are being implemented.

Beginning in 1985, the LADWP has undertaken an intensive program to rehabilitate the pre-1940 distribution pipelines to improve hydraulic capacity, water quality, and leak reduction, but important seismic improvement byproducts also result from this program. To date, the rehabilitation program has focused on using trenchless technologies to cement mortar line the pipelines, as shown in Figure 26, or insert a new high density polyethylene (HDPE) liner into the host pipe, as shown in Figure 27, because these methods are usually much less expensive than replacement. Some pipelines have been completely replaced for specific pipeline and operational needs. The majority of rehabilitated pipe has utilized cement mortar lining techniques and in some years more than one million linear feet of pipe was lined utilizing at least four lining contactors the year around.

Figure 26. Example of steel pipeline before and after cement lining.

Figure 27. HDPE lining of 36-inch (900 mm) diameter Pico Boulevard riveted steel pipe.

An additional benefit obtained from the distribution pipe rehabilitation is the seismic mitigation inherently achieved through the necessary process of performing the lining installation, which requires all the old valves, fittings (elbows, tees, crosses, and bends), fire hydrant laterals, and service connections to be removed and replaced with new appurtenances. The HDPE also provides flexibility to the existing rigid pipe joints;

most of the unlined pipe was installed with rigid lead or cement joints. The pipelines are also inspected at the exposed locations to determine their physical condition. In essence an almost new pipeline has been created using trenchless technology, the cost of which may be as low as one-third the cost of a new pipeline.

Trenchless technologies have also been used on LADWP large diameter supply trunk lines to improve water quality, reduce leaks, and improve seismic performance [56]. For example, the 40-inch (1020 mm) diameter Roscoe Boulevard and 36-inch (900 mm) diameter Pico Boulevard old riveted steel pipelines were improved by slip lining. This entailed placing a new, slightly smaller diameter, HDPE pipe into the annular space of the existing pipelines (Figure 27). Other LADWP trunk lines have been and are planned to be improved using this method.

STORM WATER/FLOOD

Storm water exposes the Water System to flooding and water quality hazards. The early reservoirs in the Santa Monica Mountains constructed by William Mulholland for normal operation and emergency storage were placed in canyons with no urban development and thus with very little impact from the watershed on water quality. As reservoir watershed development occurred, storm water channels, diversion facilities, and retention basins were constructed. As examples, Silver Lake and Hollywood Reservoirs have storm water channels which have been incorporated into roadways with high water diversion walls. Flooding along the LAA has caused damage to pipes and channels at canyon crossings where the LAA has been hardened for improved resistance; the following Debris Flow section identifies additional related LAA damages.

In the Van Norman Complex an extensive set of facilitates were constructed in 1940 to divert storm water around the Upper and Lower Van Norman Reservoirs, including detention basins, channels, and storm water pipes [9]. North and west of the reservoirs a network of four debris retention basins connected by large diameter buried pipes and channels were built to divert the waters of Weldon Creek, Bee Creek, Grapevine Creek, and San Fernando Creek around the west side of the reservoirs. On the east side the Lakeside and Yarnell Debris Basins were constructed. The Los Angles Reservoir was placed into service in 1979 to replace the Upper and Lower Van Norman Reservoirs. All storm water is routed around the Los Angeles Reservoir through the existing storm water diversion system established for the Van Norman Reservoirs, an additional storm water detention basin north of the reservoir, and a channel around the reservoir's east side. The Upper and Lower Van Norman Reservoirs were converted into retention basins benefiting the downstream urban development.

Since the enactment of the Surface Water Treatment Rule (SWTR) by the Federal and State governments, all open distribution reservoirs must eliminate all surface water from entering the reservoir or provide full water treatment prior to serving the customers. This has caused extensive revision in the City's distribution system, removing several large storage reservoirs from providing supply for normal distribution, and has downgraded several open reservoirs to function only as emergency storage facilities, such as Hollywood, Encino, and Stone Canyon Reservoirs. Smaller reservoirs, such as the Ascot, Highland, and Rowena Reservoirs, were replaced by tanks; some of these reservoir replacements were a result of enacting multihazard concerns considering seismic, surface water treatment rule, and other issues.

DEBRIS FLOW

The primary debris flow hazards are along the LAA, which block channels and fill reservoirs, but some debris flows have affected the Water System in Los Angeles. The LADWP installed crossings as shown in Figure 28 over the LAA open channels (overcrossings) where the LAA was intersected by creek beds. Most of the crossings are concrete bridges as shown in Figure 28, but some are half pipe sections, that allow water and debris from the creeks to flow over the LAA channels. Prior to implementing a crossing improvement program in the 1970's, a significant volume of sediment debris built up in the North Haiwee Reservoir and blocked LAA flow into the reservoir. The rate of sediment deposition and blocking of the LAA was significantly reduced following the additional overcrossing installations.

Another recent example of the crossing effectiveness, which also shows what happens in the absence of the overcrossings and when the overcrossings are over-inundated with debris, occurred on July 31, 2003 during a flash flood. The flash flood occurred west of the Owens Lake and caused a large volume of sediment debris to flow down several creeks. Several overcrossings in the flash flood area were able to control the debris flow, several feet in depth, over top of the LAA channel. However, as shown in Figure 29 one overcrossing for a small normally dry creek extending out of a very small canyon was so saturated with rainwater that the flash flood waters inundated and deposited enough debris to completely plug the passage over the LAA, resulting in a significant volume of sediment depositing in the LAA channel. The 12-foot (3.7 m) deep and 30-foot (9.1 m) wide top LAA trapezoidal channel was completely filled with sediment for a length of approximately 2,000 feet (610 m). The sediment completely plugged the LAA, causing it to overtop and spill water at a nearby emergency spillway, which then flowed downhill and flooded Highway 395 causing a temporary road closure. The failure of this single crossing is a very rare event due to an unusual rainfall condition in a small canyon. However, this example shows the importance of preventing debris flow from entering the LAA and how the crossings normally prevent sediment from entering; all but one overcrossing performed well. Other similar debris flow problems have occurred along the LAA in the past.

Figure 28. Typical overcrossing of LAA.

Figure 29. 2003 debris flow inundating overcrossing and LAA west of Owens Lake.

Flooding and debris flows have occurred along the LAA in this manner since it was completed. Catherine Mulholland [1] identified significant damages in June 1925 where storms washed out 400 feet (122 m) of LAA near Jawbone Canyon, Sand Canyon, and Dove Springs and tons of debris clogged the LAA preventing water flow to the City. This damage took more than 4 weeks to repair, requiring the City to survive on local water storage. In September 1927 cloud bursts clogged portions of the LAA. These are only two of many examples of this type of problem that the overcrossings help mitigate.

An example of debris flow in the City of Los Angles comes in 1995 when the Upper and Middle Debris Basins (UDB and MDB) on the VNC became inundated with sediment during large El Niño storms [57]. The debris resulted from erosion of landslide materials in the surrounding Santa Susana Mountains tributary watershed, which existed as a result of the numerous landslides that occurred during the 1994 Northridge Earthquake [58]. The UDB and MDB are relatively large basins, covering a total of 18 acres (7.3 ha), with a large tributary watershed, that were constructed in 1940 for storm water and debris control as described in the previous section. The basins served their purpose in collecting the sediment; however the rapid inundation of debris threatened important water operations from the adjacent Metropolitan Water District of Southern California's Joseph Jensen Filtration Plant by blocking the emergency overflow pipes requiring emergency construction activities. Sedimentation also affected water flow in the basins and with the continuation of large El Niño storms erosion of power transmission tower foundations and the surrounding embankments became a problem. The embankments along the basins protect storm water from flooding important Los Angeles Water System facilities and also support critical water, gas, and crude oil pipelines; damage to these facilities could result in severe secondary disasters. Thus, the combination of multiple hazards, the 1994 earthquake, landslides, and the 1995 El Niño storms, created a debris flow that was controlled with existing facilities but required emergency operations and extensive construction activities to remove the debris to protect continued operations and other critical lifelines.

DROUGHT

The City of Los Angeles is located in a semi-arid region that is periodically exposed to drought, which impacts the ability to provide a safe and reliable water supply. As a result, the LADWP is very proactive in implementing measures to reduce drought affects. The development of several supply sources, including the LAA, and storage reservoirs as described at the beginning of this report were to help reduce drought impacts. The LADWP played a primary role in planning for building the Colorado River Aqueduct and Hoover Dam, both constructed at sites and alignments originally favored and proposed by Mulholland [1]. Other measures include water conservation, water recycling, and developing alternate water supplies [59].

Conservation

The LADWP was a pioneer in conservation by requiring water meters to measure the volume of water used by customers and discourage excessive use by charging for the water. In 1925 the LADWP became 100 percent metered.

The LADWP implements one of the most innovative and successful water conservation programs in the nation, resulting in a demand reduction of more than 15% annually for more than a decade. Program efforts have focused on residential customers where more than 1.2 million standard toilets, using up to 7.0 gallons (28.4 liters) per flush, have been replaced with ultra-low-flush (ULF) toilets, using only 1.6 gallons (6 liters) per flush. This has resulted in an annual water savings of more than 11 billion gallons (42 billion liters). Efforts are being made to expand the ULF program to commercial, industrial, and institutional customers.

In addition to the ULF toilet program, rebates are offered for the installation of ULF urinals, cooling tower conductivity controllers, residential and coin operated high efficiency clothes washing machines and incentives for large business custom water conservation projects. Free two-gallon (7.6 liters) per minute water saving showerheads are available to residential customers. Other conservation activities include a residential landscape irrigation program and installation for pre-rinse spray heads in restaurant and other food service establishments; numerous financial incentives are provided by the LADWP to implement these conservation programs. A two-tier water rate structure has been implemented to encourage water conservation; a lower rate for basic needs up to a defined quantity of use and a higher rate for other needs exceeding the base level.

Water Recycling

The San Fernando Valley Water Recycling Program is proposed to provide recycled water for irrigation, commercial and industrial use from treated water from the Donald C. Tillman Water Reclamation Plant. Reclaimed water from the West Basin Water Reclamation Plant provides recycled water for the Westside Water Recycling Project, which provides water for irrigation and industrial use in the Los Angeles International Airport area. In the Port of Los Angeles area water from the Terminal Island Wastewater Treatment plant will be used to provide recycled water for industrial and the seawater intrusion barrier. The Los Angeles Greenbelt Project has been in operation since 1992 providing water for Griffith Park and commercial customers for landscape irrigation. Funding for these projects comes from federal, state, and local sources.

Alternative Water Supplies

Water conservation and recycling provide near-term water supplies. The LADWP is looking at a number of alternative water supplies to meet the need for future demands. Water marketing, the transfer lease or sale of water or water rights, allows the transfer of water without building costly facilities. However, many issues including economic, environmental and third party, must be addressed because of the unique nature of water. Desalinating seawater has been technically feasible for a number of years; however, the costs have not been comparable to conventional water sources. The LADWP is participating with several demonstration and research projects to refine the process and cost of desalination. In the 1920's, William Mulholland initially implemented conjunctive water use in the San Fernando Valley Groundwater Basin to store imported water supplies in groundwater basins and reservoirs for later provision in times of drought. This allows the recharging of basins during wet years and water extraction by pumping in dry years.

FIRES

The City of Los Angeles is subject to urban structural fires and wildfires in reservoir watersheds and the open mountainous brush ranges. The design of the City's water distribution system, in addition to providing for customer residential, commercial, and industrial water demand, provides capacity for fire suppression supply from the City's fire hydrants and industrial customer-owned internal fire suppression and fire sprinkler systems. The City's pump-tank systems have been design for fire suppression by including additional capacity in the tanks for fire supply storage and additional pumping units for fire. These units are usually internal combustion (IC) units in case there is electricity outage. In some cases the IC units drive electric generators.

In the 1960's the LADWP and Los Angels City Fire Department (LAFD) (Earl Leonard and the lead author) identified and documented a number of locations where fire engine pumping units can be located to allow the unit to pump from a lower-pressure zone to a higher-pressure zone through adjacent fire hydrants [60]. Pumping for fire suppression was utilized following the 1971 San Fernando and 1994 Northridge Earthquakes and other emergencies where the system sustained temporary loss of supply.

Helistops have been provided in mountainous regions such as Elysian Park and at reservoir sites (for example Santa Ynez Reservoir) for LAFD use to operate for suppression of brush fires. At some locations fire hydrants have been provided for the water-dropping helicopters to refill their tanks. "Super Scoopers", large water dropping fixed wing aircraft which can fly low over a large reservoir and scoop up water to fill their tanks, have utilized Silver Lake and Bouquet Reservoirs for fighting fires.

In reservoir watershed areas firebreaks were installed where ever possible. In the Stone Canyon Reservoir watershed heavy drums filled with water have been rolled up and down the slopes to compact the brush in order reduce the height of the flammable brush. High brush when ignited tends to blow into the air, forming a chimney effect, and travels to other areas including the surrounding housing developments.

LANDSLIDES

Landslides are of great concern and are one of the most common hazards for the Los Angeles Water System. Slope movements cause damage to supply and distribution pipes, reservoirs, and other facilities located in the mountains and along slopes. Landslides occur often and the Water System maintains a staff of engineering geologists and civil/geotechnical engineers to help mitigate their effects. The listing of landslide effects on the Los Angeles Water System and the mitigation methods utilized are too extensive for the scope of this report and therefore descriptions will be limited to generalized problems and mitigations and some specific cases.

Landslides are affected by the influence of water, which can weaken the shear resistance along the sliding plane, increase the driving forces through saturation of the slide material, etc. Thus, the business of storing and distributing water creates an inherent conflict with the landslide hazard in regions of pre-existing steep and unstable slopes such as those existing in Los Angeles. One common difficulty is where slope movements cause a water pipe to leak and the leaking water leads to further slope movements, or an old leaky pipe helps initiate movement. Reservoirs can also have some problems in less favorable geologic areas by saturating the reservoir side slopes which can weaken potential slide planes and cause movements. The most effective landslide mitigation is to be proactive in identifying the potential hazard and avoid the hazard wherever possible. Avoidance is not always possible, so improvements to limit or eliminate landslide effects are sometimes necessary. Sometimes landslide problems are not identified and remediation following an occurrence is necessary.

Some common methods utilized by the LADWP to mitigate landslides are: install buttress at base of slope; place retaining walls, crib walls, or similar structural or gravity retaining structures; remove and reconstruct with compacted earth fill or soil cement to a stable slope inclination; excavate head of slope to remove driving forces; use structural (e.g., piles) and/or soil improvements (e.g., stone columns) to increase shearing resistance across failure plane; install tie-back rods and anchors; install subdrains and horizontal auger drains to remove groundwater; provide surface drainage to divert water away from slope to protect it from saturation and surface erosion; reduce landscaping irrigation. Structures at the top of slope that may potentially be subjected to limited movements (e.g., seismically induced) have been constructed with a layer of compacted fill below the foundation to eliminate structural damage or reduce it to controllable levels. Instrumentation for monitoring landslides includes slope indicators, movement and settlement monuments, and strain gages to help understand and predict slope movements.

Additional mitigation methods used to control movements on reservoir slopes include: place blanket drains to control seepage into and out of slopes; place pervious asphalt liner to allow drainage when the reservoir level is lowered; control rate of lowering reservoir level to prevent pore pressure buildup in slopes resulting from a rapid drawdown condition.

Some methods to mitigate slope movements on pipelines are: remove and relocate pipes out of slide areas; replace buried pipe in slide to an above ground pipe over slide (sometimes called sidelining); install flexible mechanical couplings on pipes passing though deformation regions; use solid connected pipes, such as welded steel or other restrained joint pipe, unlikely to leak under deformation; install isolation valves; leave loose backfill around buried pipes in slide zones.

One of the first examples of a serious landslide affecting a Los Angeles Water System facility comes from the St. Francis Dam failure. The concrete gravity dam was unknowingly founded on a large paleolandslide comprised of adversely dipping Sierra Pelona Schist. This failure, shown in Figure 5, apparently resulted from a sequence of events that included movement of the ancient megaslide on the left abutment [2]. The remnants of the slope movements can be seen on the right side of Figure 5.

Figures 30 and 31 show landslides that occurred in Upper Stone Canyon Reservoir. Grading for the reservoir's eastside undercut adversely dipping shale beds and possible old landslide planes located along bedding, causing the landslide shown in

Figure 30. Filling of the reservoir allowed water to enter the bedrock comprising the landslide area. In 1956, 51,300 cy (39,240 m^3) of material slid into the reservoir. Over 171,000 cy (130,800 m^3) of landslide material above and in the reservoir was removed. Unfortunately, the head of the graded area undercut bedding and slide planes above the 1956 landslide. After unusually heavy rainfall in 1969 another large landslide occurred above the 1956 landslide. The 1969 landslide did not enter the reservoir [61].

Figure 30. 1956 landslide in the Upper Stone Canyon Reservoir.

Figure 31. 1956 rapid drawdown landslide in the Upper Stone Canyon Reservoir.

The 1956 landslide shown in Figure 30 was remediated by drilling several horizontal drains into the hillside for dewatering. Also after grading out the slide, a relatively impervious fill blanket with an underlying extensive subdrain system was installed below the reservoir liner in order to limit and intercept any reservoir water before it enters the hillside in the landslide area. The 1969 landslide was removed by grading. The entire slope was trimmed to a lower angle than the bedding dip.

Figure 31 shows the 1956 failure along the Upper Stone Canyon Reservoir interior slope that occurred along reservoir west side as a result of lowering the water elevation too rapidly after the occurrence of the landslide shown in Figure 30. The failure involved bedrock bounded by fault gouge that became unstable because the reservoir was drained so rapidly that water pressure in the bedrock did not have time to drain and reach equilibrium. The slope failure was removed by grading. An 8-foot (2.4-m) wide fill blanket was placed in the slide area and horizontal drains were drilled into the slope. A 6-inch (150-mm) diameter open tile drain was placed at the toe of the slope and a fill blanket was placed along the entire west side of the reservoir. Additionally, it was recommended that all reservoirs be drained at a slower rate to allow bank storage water to escape and decrease pressures [61].

VOLCANISM

The Mono Basin and Owens Valley, where the LAA originates, are located in a volcanically active region. A volcanic eruption would threaten water system facilities and a major source of Los Angeles water supply. The LAA provides approximately 50% of the annual water supply for City of Los Angeles. Most of the LAA water (60% of the aqueduct water; 30% of the total supply for the City) originates form the Long Valley Caldera and Mono Craters Area north of Bishop California. The Long Valley Caldera is approximately 10.5 mi (17 km) wide by 18.5 mi (30 km) long and was formed 760,000 years ago by a catastrophic volcanic eruption 2,000 times the size of Mount St. Helens. The entire caldera has had hundreds of small to medium sized eruptions since, with the latest occurrence 250 years ago. Since 1978 markedly increased seismic activity related to volcanic activity (up to 1,000 earthquakes per day), ground surface deformations and increased carbon dioxide emissions provide strong evidence that the area is still active and capable of renewed volcanic eruptions. However, at this time it is unknown where and when the next eruption will occur. Hot springs and geothermal areas occur throughout the Owens Valley, but the primary volcanic hazard is the Long Valley Caldera.

The LADWP has prepared a contingency plan identifying the emergency response measures to be taken in the event of a volcanic eruption [62]. The plan accounts for protection of personnel, facilities, and water supply, and also considers secondary disasters associated with dams and reservoirs. If concern of a volcanic eruption were to arise the Water System plans to evacuate personnel, stop imports of water north of the eruption area and increase water flows south of the eruption area to maximize protection of supply and minimize the potential for secondary disasters. Alternate sources of supply will be utilized, such as the MWD and groundwater, and water conservation actions may be implemented.

SOIL SETTLEMENT

Ground settlement has affected water conduits causing concern for leaks. In the 1950's, the Eagle Rock-Hollywood Conduit, large diameter pipeline, was constructed to connect Eagle Rock Reservoir with the Upper Hollywood Reservoir to provide MWD water to the Hollywood-Wilshire area and provide redundancy of supply in that area. Along Franklin Avenue at Talmage Street the conduit route passed through an old trash dump, consisting of tin cans and other household debris. Monitoring of the 68-inch (1,725-mm) reinforced concrete cylinder pipe with rubber gasket joints was done and after a few years subsidence began to occur. Remediation of this subsidence was accomplished by installing a welded steel liner within the existing pipe. This created in effect a double conduit in this area. Seepage and settlement monitors were installed.

The Mojave Conduit, a gravity reinforced concrete box, was constructed for the Second Los Angeles Aqueduct (SLAA) in the northern Antelope Valley, north of Mojave. The conduit crossed the alluvial fan of Cache Creek runoff from the southern Sierra Nevada. The material deposited was not completely consolidated and subsided when subjected to additional water. The conduit was constructed with joints that had been sealed, but were not completely watertight. After several years of the operation of the SLAA water seeped from the joints and began to consolidate the alluvial material and the conduit settled. The joints were resealed; however there was the necessity to return the conduit to its original hydraulic grade. Injecting mud (bentonite) under the conduit was done with care; increasing the mud pressure brought the conduit back to design gravity grade, then the joints were sealed one final time.

COLD WEATHER

Cold weather exposes the LAA and other northern facilities to problems that are not found in the warmer climates of Los Angeles. The LADWP constructed the Monolith Cement Plant in Tehachapi California to provide cement for LAA construction. This plant developed and manufactured a Modified Type IIA cement which has been used by the LADWP since that time because it provides freeze-thaw protection by use of air entrainment. This cement meets specifications for ASTM C-150. Freeze-thaw and snow conditions require more maintenance of facilities than those not exposed to such conditions. For example, the concrete spillways surfaces for Long Valley and Grant Lake Dams were deteriorating under cold weather conditions; the spillways were repaired by pressure washing the surfaces placing a new concrete overlay at Long Valley and using a durable Polyurea coating for Grant Lake. The Polyurea coating was also used on a portion of Long Valley spillway and provides protection to continued cold weather deterioration.

The Second Los Angeles Aqueduct broke in January 1970, as shown in Figure 32, at two points in an above ground pipe location in Cemetery Canyon, 21 mi (34 km) north of Mojave, CA and in January 1971 in a buried section 17 mi (27 km), near Cinco, north of Mojave when exposed to an air temperature of $9°$ F ($-13°$ C) (installation temperature was $75°$ F, $24°$ C) [63]. The pipe was 82 in (2100 mm) diameter, 15/16 in (23.8 mm) thick, under a hydrostatic head 1,085 ft (331 m), shop fabricated with ASTM-572, Grade 42, Type II high strength, low alloy steel, in 40 ft (24 m) lengths, and then shop welded into 80 ft (24 m) lengths; field welds were every 80-feet (24 m).

All three breaks occurred in the bell of a bell and spigot joint at the toe of the inside fillet weld. Cemetery Canyon had double welded field joints and the Cinco failure was a single inside shop-welded joint. The unique nature of these failures led to considerable research to determine their causes, and identified that all three breaks resulted from brittle fracture developed by triaxial tension; the two failures in Cemetery Canyon were aided by improper welding that left a notch in the base metal and the one at Cinco was brought forth by development of a shallow crack. In using the high strength, low alloy steel, the gain in strength is at the price of lower ductility, and the metal becomes more brittle with lower temperature. The research performed by the LADWP identified that the state of practice for welding high strength alloy steel was inadequate for cold weather conditions, and the problem was compounded by the additional stresses developed in bell and spigot joints.

Figure 32. Second Los Angeles Aqueduct 1970 break in Cemetery Canyon [64].

The joint welding condition was mitigated by installing expansion joints, implementing corrective welding procedures, and replacing some bell and spigot joints with butt-welded joints to reduce tensile loading, reduce stress concentrations, and increase joint strength. In addition, overall improvements were made to the LADWP welding design, specification, and inspection techniques, for normal and high strength steels, as well as improvements in specifying high strength steel requirements for meeting toughness requirements at specified low temperatures. The Water System also reconsidered its philosophy for the use of the bell and spigot vs. butt strap joints.

WIND AND SEICHE

The design of LADWP reservoirs provides for adequate freeboard to prevent overtopping by wave action or seiche and normal crest settlement. The dam upstream slopes and many of the reservoir side slopes are paved to help keep waves from eroding them and for stability and water quality problems.

The transport of Owens River water to Los Angeles has lowered the Owens Lake water levels. When high winds blow over the dry lake bed, large dust clouds develop and create poor air quality. The Water System is implementing measures, costing several hundred million dollars, to mitigate the air quality from the Owens Lake through rewatering, planting, irrigating, and placing coarser materials.

WATER QUALITY

Water quality is a significant issue and the Los Angeles Water System has always maintained a high priority to ensure long term health effects and disasters resulting from water born disease or pollutants never arise. Some water quality issues have previously been discussed in relation to other hazards and will not be repeated here. With the exception of Cross-Connections described below, detailed descriptions of potable water quality are beyond the scope of this report and interested readers are referred to other technical literature (e.g., [65][66][67]). The Water System also implements measures to reduce urban storm water runoff water quality impacts, but these are considered beyond the scope of this report because local storm water quality impacts are generally not hazardous to the City's domestic water supply. The Los Angeles Water System has been and continues to be proactive in water quality issues and maintains a large organization to deal with these matters on a daily basis. In addition to those previously identified, the following is a brief general summary of the extensive history.

The LADWP has met or surpassed all drinking water standards set by the United States Environmental Protection Agency (EPA) and the State of California Department of Health Services (DHS). This accomplishment is largely due to application of state-of-art water treatment processes, prudent facility management and operation and vigilant monitoring and testing of the water quality constituents. The LADWP annually collects more than 25,000 samples and conducts more than 200,000 tests in its own modern laboratory and other state-certified laboratories.

In the watersheds of the Owens Valley and Mono Basin, restrictions on the use of certain activities on lands leased from the City that might affect the quality of the water supply are included in any land use agreements. For example non-EPA authorized harmful chemicals are not permitted on lands leased for agriculture. Some of the lands leased for cattle grazing are fenced off from streams to prevent erosion damage to the fishery.

To comply with the Disinfection By-Product Rule, the LADWP completed a system-wide evaluation of converting the water disinfection process from free-chlorine to chloramines [68]. The evaluation concluded that chloramine conversion will reduce the level of disinfection by-products to a much lower value than the established standards and will also provide flexibility and increased reliability of supply with the purchased chloraminated water from MWD. Also, the LADWP has embarked upon a feasibility study of further reducing arsenic from the surface water supply by enhanced coagulation at the Los Angeles Aqueduct Filtration Plant. In this study, in addition to arsenic, other water quality improvement benefits of enhanced coagulation are being investigated.

Cross-Connections

Connections providing water service to industrial, commercial, and landscaping are sources of pollution from the potential of contaminant backflow through the water meter during a loss of pressure in the domestic water system. The LADWP was a pioneer and leader in the recognition and prevention of this potential contaminant source. In 1932 LADWP engineers develop a cross connection control device, which consisted of two standard swing check valves and a relief valve in series. Prior to that time there were little or no backflow protection. The metal clappers in the check valves were modified with a rubber disk to provide a watertight seal. Figure 33 shows 2 configurations for modern cross connection control devices.

In 1945 the LADWP created the Foundation for Cross Connection and Hydraulic Research with the University of Southern California, led by LADWP Senior Sanitary Engineers Ray Derby and Roy Van Meter, to develop backflow prevention criteria for the manufacture and testing of the backflow prevention valves. The Foundation developed initial criteria for the valves [69] and later updates under contract with the LADWP [70]. A manual for cross-connection control was developed by a four-man committee [71], two of which were LADWP employees. The laboratory was originally located on the USC campus. In 1953 the LADWP provided a building and facilities at the former Riverside Pumping Plant for the Foundation research, training, and the testing of manufactured back flow devices. The laboratory continues to function at this location.

Figure 33. Two styles of modern backflow prevention devices.

Cross-connection control operations for installation, testing, and maintenance are presently required to conform to the California State Cross-Connection Regulations, Title 17, of the California Administrative Code and the LADWP Rule 16-D. Enforcement for testing and maintenance is provided by the Los Angeles County Department of Health Services.

HAZARDOUS MATERIALS

Hazardous materials pose risks for (1) the protection of the City's water supply against a potential material release and (2) the managing and handling of chemicals needed for water quality and operations that may result in health and fire hazards. Examples of protecting the water supply from hazardous materials are provided for the LAA in the Earthquake - Present and Future Perspectives section of this report and for some water storage facilities as a byproduct of covering reservoirs for water quality purposes described in various sections of this report; other examples exist but will not be presented here. This section mainly focuses on hazardous materials needed for water quality and operations.

Most chemicals utilized by the Water System are for the control of water quality, some are for the general health improvement of the public (e.g., fluoridation), and others are simply for general operation (e.g., fuels, solvents, pesticides, etc.). The chemicals used for water quality include chlorine, sodium hypochlorite, chloramines, ozone, sodium fluorosilic acid, zinc-orthophosphate, ferric chloride and cationic polymer. The specific purpose and use of each chemical and descriptions of their hazards to people and the environment are beyond the scope of this report. The above listing shows that the LADWP handles quantities of hazardous materials that require adequate management for the protection of Water System employees, the general public, and the surrounding environment.

Hazard materials management is regulated by federal and state codes. The handling and storage of hazardous chemicals within the City are coordinated closely with the Los Angeles Fire Department, who is the designated regulating enforcement agency for the City. City regulations include spill mitigation and containment and securing hazard materials containers to prevent spills. Certain chemical storage and handling require approvals from other agencies such as the California Department of Health Services. The LADWP hazardous material storage and handling meet regulatory requirements. The design of chlorine facilities follows the Chlorine Institute and National Center for Disease Control guidelines. Risk management and prevention plans are prepared for hazardous materials storage facilities.

Manufacturers of chemicals are required to prepare Material Safety Data Sheets (MSDS) for all products delivered to their customers. The MSDS are required to be posted at all locations where chemicals are stored or used and identify the hazardous ingredients; physical properties; reactivity data; spill, leak, and disposal procedures; special handling information; fire and explosion; health hazard information; and first aid. The LADWP trains appropriate personnel in the use and storage of all chemicals in use. In addition, employees and contractors who do not handle chemicals, but must work on sites containing hazardous materials such as chlorine, are trained for the hazard exposure prior to entering the site.

DAM AND RESERVOIR SAFETY

Water retaining dams pose a significant potential hazard to downstream life and property. The Los Angeles Water System maintains many dams for the primary purpose of water storage, but a few serve for storm water control. The number of in-service dams has changed over time, and some of the dams have changed their purpose and function since originally constructed. All of the Los Angeles Water System dams having a minimum height of 25-feet high (7.6 m) or retaining at least 50 acre-ft (61,674 m^3) volume of water fall under the jurisdiction of the California Department of Water Resources Division of Safety of Dams (DSOD). At its peak the Water System had 30 jurisdictional operating reservoirs, an additional 8 non-jurisdictional operating reservoirs, and 3 reservoirs for storm water control. At present there are 30 reservoirs under DSOD jurisdiction, of which 23 are operating reservoirs, one being retained by a concrete gravity dam and all others by earth fill embankment dams; several reservoirs are retained by multiple dams.

The LADWP has always considered dam safety important, and even employed the advice of eminent dam engineers and geologists in the early 1900's during the design and construction of large dams [2]. However, constructing large dams in an era of developing dam technologies and owning and operating dams in the unique geologic environment of Southern and Eastern California have had significant consequences for the Los Angeles Water System. The St. Francis Dam failure, shown in Figure 5, catastrophically released a large volume of water causing serious damage to downstream property and hundreds of people lost their lives [4]. The Baldwin Hills Reservoir failure in 1963, shown in Figure 34, released reservoir water that damaged property and five people indirectly lost their lives. The LSFD upstream slope failure in 1971 [11], shown in Figure 13, did not release water but was a close call requiring evacuation of at least 80,000 people. The St. Francis and LSFD failures were previously discussed in the Earthquake and Landslide sections and will not be repeated here.

The Baldwin Hills Reservoir was constructed in 1951 as a regulating reservoir to serve southwestern portions of Los Angeles. The design considered the Newport-Inglewood fault system and location of the Inglewood oil and gas field. A very sophisticated monitoring system was installed recognizing the potential hazards. A ten foot (3 m) thick compacted earth fill blanket was placed on the reservoir slopes and bottom to prevent seepage and a drainage system was installed beneath the blanket to detect seepage concerns. In 1955 the oil companies began a program to inject fluids into the ground under pressure for secondary oil and gas recovery from the Inglewood field. The volume of injected fluid increased by about a factor of four in 1960, which led to the rupture and offset of Newport-Inglewood fault splays extending under the reservoir and through the dam's right abutment. The ground offset provided space for water seepage to erode natural material, which led to the abutment failure shown in Figure 34. The reservoir monitoring system worked to provide an early warning of the upcoming failure. On December 14, 1963 the reservoir caretaker detected increases in seepage flow and alerted LADWP supervisors. The Los Angeles Police Department evacuated downstream of the reservoir shortly before the reservoir failed and released water into the downstream residential neighborhood. The early detection saved many lives [72][73]. This disaster provided a valuable lesson for the Los Angeles Water System to not have reservoirs in active oil and gas fields, and the lesson has been heeded several times by denying oil companies the permission to extract oil and gas from below the several dams located on the Van Norman Complex, shown in Figure 14.

Figure 34. 1963 Baldwin Hills Reservoir failure through the dam abutment [73].

The primary outcome of the St. Francis Dam failure was the creation of dam safety legislation enacting a new State dam safety program that is now administered by the DSOD to ensure dam owners focus attention on dam safety issues and reduce chances of such dam-related catastrophes. Initially the DSOD held jurisdiction over dams located within a natural water course. After the Baldwin Hills Reservoir failure in 1963, a LADWP off-stream reservoir, the legislature increased the DSOD jurisdiction to cover off-stream reservoirs. The LSFD upstream slope failure was cause for great concern in dam safety and initiated a program for evaluating seismic stability of all jurisdictional dams. These three dams represent the most notable dam safety issues in California [74] and studies of them have provided many lessons learned resulting in improvements in dam and reservoir safety for the LADWP and other dam owners. These three dams were each constructed using the latest technologies of their time, but their failures identified improvements necessary to help mitigate hazards that exist with large reservoirs.

Mitigation strategies resulting from these failures include improved evaluation, design, construction, and monitoring methods; the most significant of which is the development of modern soil compaction and testing methodologies and numerical modeling techniques described earlier in the Earthquake section this report. Other emergency preparedness mitigation strategies resulting from studies of these dams include downstream flood evaluations resulting from a dam break and development of inundation zone maps, pre-planned evacuation procedures, emergency blow-off procedures, etc.

The Los Angeles Water System maintains a Reservoir Surveillance Group of engineers and technicians who regularly inspect and continually monitor dam and reservoir data. This group interacts with operating and design personnel, and together they all make up the core components of the Water System dam safety program. Dam analysis and design are performed and administered by design personnel and the Reservoir Surveillance Group reviews for dam safety elements and monitoring. All system, facility, and maintenance modifications affecting dams and reservoirs are also reviewed for dam safety. In addition, construction projects, dam modifications, or other facilities encroaching on a dam, are closely monitored for conformance with dam safety and operations criteria. Engineering studies performed for the dams include stability (seismic, rapid drawdown, etc.), reservoir blow-off and drawdown, flood hydrology, spillway capacity, and seepage. In addition, inundation maps are prepared for each dam. The reservoir outlet tower gates and valves used to blow-off the reservoirs are exercised twice per year. The Reservoir Surveillance Group maintains emergency preparedness plans and equipment and responds to emergencies when needed, as described in the Emergency Preparedness and Recovery section of this report.

Operating personnel inspect most dams and reservoirs daily, however some are inspected weekly. The Reservoir Surveillance Group is based in Los Angeles and inspects most distribution and southern aqueduct dams monthly, with a few low-hazard dams inspected quarterly, semi-annually, or annually; the more distant northern aqueduct reservoirs are inspected annually. These inspections are documented and maintenance recommendations forwarded to operating personnel. A listing of needed dam safety maintenance activities are monitored through completion. Detailed engineering inspections are performed approximately once per year and a report prepared to describe the findings. In addition, physical inspections are performed and reported for each jurisdictional dam by DSOD and LADWP engineers at least once per year. A large variety of special inspections of spillways, drainage systems, instrument calibrations, seismic equipment maintenance, slope indicators, etc. are conducted on a periodic basis.

Seepage monitoring systems at many dams send continuous data to the Los Angeles Water System Data Acquisition System Center (LAWSDAC), a Supervisory Control and Data Acquisition (SCADA) system. Seepage at all monitoring points on dams is measured weekly and piezometers are read monthly. Seepage samples are periodically collected and measured for total dissolved solids. Surveys of all dams are performed from one to three times per year to precisely monitor horizontal movement and settlement. A database is maintained for all surveillance data, which is monitored and processed. Reports are prepared, approximately annually, and sent to the DSOD. Concerns arising from surveillance data are initially investigated by the Reservoir Surveillance Group, and more detailed and advanced studies are performed by engineering groups and consultants when necessary.

Several examples of dam safety concerns and problems were previously described and will not be repeated here. Seepage problems at the LSFD, Harbor Reservoir, and other dams required extensive grouting programs to control seepage flows and pressures and internal erosion; these dams reservoirs are no longer in service. Other means for controlling seepage concerns is to impose reservoir elevation restrictions. Water distribution system and reservoir site modifications that are undertaken for other hazard improvements, such as water quality, require thorough review of potential impacts on dam and reservoir safety issues, and in some cases additional dam and reservoir mitigations are necessary to implement as a result of other hazard mitigations. Reservoir site construction modifications also require more stringent safety monitoring.

HOMELAND SECURITY

The LADWP is currently implementing measures to prevent potential terrorist threats against the Water System in the post-9/11 era. Historically the Water System has had to ensure security and enact counter measures against attack during the 1920's "Water Wars", World War I (WWI), World War II (WWII), and the cold war. In addition, the Water System responded to system damages inflicted during the 1965 (Watts) and 1992 (Rodney King) civil unrest in Los Angeles. Security during these civil disturbances was mainly from police protection for employees performing system repairs, following sabotage, to ensure an adequate water supply for fire fighting. Due to the limited mitigation efforts the civil disturbance hazard will not be discussed further.

Post-9/11 Terrorism

Following the terrorist attacks on the World Trade Center in New York and the Pentagon in Washington D.C. on September 11, 2001, the LADWP has increased security and sought to reduce the potential for terrorism to Water System facilities. Expert consultants familiar with terrorism and security issues have been retained to advise the Water System on measures to be implemented. Risk and vulnerability assessments have been performed for critical facilities and countermeasures identified. Some general measures implemented to reduce risk and vulnerabilities include: increase in security personnel, hardening facilities with capital improvements, increase surveillance of critical facilities (including the research and installation of electronic surveillance equipment), place barriers around critical facilities, enforce strict facility access requirements, and increase water quality monitoring. Helicopter surveillance flights are performed over critical water and power facilities. The LADWP has also worked closely with various law enforcement agencies, including the Federal Bureau of Investigation, the Los Angeles County Sheriff's Department, and the City of Los Angeles Police Department Counterterrorism Units. The Water System recently obtained a rate increase in part to help cover the increased security costs. As a part of this type of hazard mitigation, more specific details of measures taken will not be described.

The system redundancy, isolation capabilities, and other factors making a resilient system previously described, also helps to make the Los Angeles Water System more secure. For example, following any potential attack that may affect limited portions of the large distribution system, similar to a large earthquake which damages only portions of the City, the Water System can isolate portions of the system following an attack, limit the affected area and continue providing reliable service to much of the City. Security hazard mitigation also benefits from other, completely unrelated, hazard mitigation efforts. For example, the covering of reservoirs for pre-2001 concerns naturally helps to protect reservoirs from terrorist threats, and the strengthening of certain types of facilities to resist earthquake shaking inherently reduces their vulnerability to terrorism.

"Water Wars"

The LADWP has previously faced hostile attack and domestic terrorism concerns. The "Water Wars" [75] between the LADWP and some people in the Owens Valley developed in the 1910's and 1920's to include hostile seizures, bombings, kidnappings, threats, and news propaganda activities that posed a threat to the Los Angeles water supply. This section summarizes these activities and how the LADWP responded to protect the water supply, but it is beyond the scope of this report to present the history and details that lead to violence, which are presented elsewhere [1][75][76][77][78]

During LAA construction there were several incidents of arson at the construction camps. In the 1920's there were disputes over land purchases, land prices, water rights, and other matters and some Owens Valley residents began sabotaging the LAA. The first bombings took place on May 21, 1924. Prior to this event, on July 15, 1917 there were two blowouts on the LAA which were suspected to result from dynamite being dropped into the LAA through manholes but remained inconclusive [1]. There were 6 or 7 more cases of dynamite attacks in the summer of 1924 [75].

On November 16, 1924, approximately 70 armed men seized the LAA at the Alabama Gates, released water to the Valley, and suspended water service to Los Angeles. The insurgents originally planned to keep the LAA hostage until their demands were met. The seizure ended after 5 days when an independent commission agreed to adjust the differences over water rights between the Owens Valley ranchers and the City. There were no arrests for the 1924 bombings and seizure.

Figure 35. LAA bombing damage during 1920's "Water Wars."

There was no violence documented in 1925. In April 1926 violence was reinitiated with bombs of a groundwater wells. On May 12, 1926 a dynamite blast knocked a 10-foot (3 m) hole in the LAA and in July 1926 a construction shack was blown up. The following year at least seven damage inflicting bombings occurred along the LAA between May and July 1927. Figure 35 shows LAA damage following a dynamite attack. One of these bombings included the kidnapping of two guards by four masked men. In addition, there were numerous bombs found along the LAA that did not detonate. Other damages inflicted include disabling water wells, cutting aqueduct phone lines, and dismantling police search lights. The bombings are reported to have taken place in a hide-and-seek type fashion with the saboteurs sneaking around the many guards to plant bombs. No one was convicted for any of the bombings.

Amidst the bombing raids of 1927, the Los Angeles Sheriff's Department received an anonymous phone call reporting a carload of armed men from Owens Valley going to dynamite the St. Francis Dam. Dozens of officers went to protect the dam site. No attack materialized [75].

The attacks ended in 1927 when the primary instigators and financers, bank owners in the Owens Valley, were arrested and convicted for embezzling Owens Valley ranchers' money [75][78]. The LAA dynamiting repairs were estimated to cost approximately $250,000 in 1927 with additional costs of lost water, security, investigations, intangibles, and other matters described below. No deaths directly related to the bombing attacks were reported.

Between 1924 and 1927 there were at least 15 cases where the LAA was blown up with additional attacks on other facilities [1][75][78]. Throughout this time of "Water Wars" Mulholland also received countless threats on his life, none of which he took seriously and there were no documented special protections established for Mulholland. Propaganda was also employed by the City and Owens Valley and several newspapers selected sides and further exploited the propaganda war. The dissenting Owens Valley leaders used propaganda and publicity as a weapon for creating fear in Los Angeles [78].

In response to all of the terrorist-type activities, the LADWP and other City officials employed several strategies to mitigate their effects and counter the advances including negotiation and arbitration, rewards, public and private investigations, increased police and security, increased intelligence and communication technologies, improved transportation capabilities, improving water storage capabilities, and removing unethical financing for Owens Valley hostile activities. Each topic is summarized below.

Meetings were held between LADWP and Owens Valley leaders in 1923, prior to the documented attacks, to try and settle differences and reduce tensions. Unfortunately, other events continued to intensify the problems. Negotiations were reinitiated soon after the first 1924 bombing and an impartial board of arbitrators was established to deal with land prices. Arbitration and mediation dealing more directly with the dispute followed the Alabama Gates seizure, but broke down in early 1925. A $10,000 reward for capture of the perpetrators who performed the initial 1924 bombing was appropriated by the Los Angeles City Council [1] and other rewards were offered in 1927 [75].

Los Angeles police and LADWP authorized private investigations were initiated soon after the initial 1924 bombing. No information was obtained to identify the perpetrators due to local sentiment obstructing information, which continued throughout the "Water Wars" as the local residents were sympathetic to the cause and banned together to conceal information leading to arrest. Los Angeles assigned several people to investigate the sabotage over the years with limited success. California Governor Richardson had the 1924 Alabama Gates investigated by the State Engineer. In 1927 President Coolidge ordered Pinkerton Men, U.S. private detectives, to the LAA [75].

Security and patrols were established in a variety of ways throughout the "Water Wars". Guards were added at the direction of the Los Angeles Mayor with orders to "shoot to kill" following suspicion that bombings may have occurred in 1917 [1]. Police and patrols were increased in 1926. The State militia was requested to intervene in the Alabama Gates seizure, apparently by both sides [1][75], but the Governor declined this and all other police requests throughout the "Water Wars". Following the seizure, LADWP paid for 24-hour guarding of the entire LAA length by private armed police. Following the kidnappings and bombing in May 1927, Mulholland ordered machine guns posted at every conduit, floodlights to illuminate trespassers, and horseback patrols along the entire LAA at hourly intervals. With additional attacks, six guards armed with Winchester rifles and Thompson machine guns (Tommy guns) were dispatched to No Name Siphon with orders to "shoot to kill" any suspect with no questions asked [75]. President Coolidge sent uniformed Federal agents to patrol the LAA [75]. An estimated 700 armed guards patrolled the LAA in 1927 with automatic guns and searchlights.

Reliable long distance communication was critical in responding to the attacks. In response to recurring communication problems resulting from telephone and telegraph lines being cut and damaged from bombings, Mulholland made an urgent request to the Federal Government in 1927 for short wave radio transmitters between Los Angeles and Owens Valley citing extreme dangers and threats to life and property. His message declared it "imperative to have emergency intelligence that radio alone can supply" [1]. Prior to this, in 1924, Los Angeles worked with the State Department of Transportation to improve routes between Los Angeles and the Owens Valley. Highway 395 was improved and paved and additional roads were constructed from the main corridors. Los Angeles proceeded with these improvements for multiple reasons: (1) to improve its own transportation capabilities to critical sites for normal and emergency operations, and (2) as a negotiated improvement to help the Owens Valley residents [78].

Through the 1910's and 1920's the LADWP increased reservoir storage capabilities. The Tinemaha Dam and Reservoir was built to regulate Owens River flooding and as a holding basin in the event of trouble between itself and Haiwee Reservoir, which was apparently initiated at Mulholland's request as a direct result of destructive events in the Owens Valley [1]. Other dams were built during this time, including St. Francis, under consideration of events that transpired in the Owens Valley, other hazards, and normal distribution needs. Completion of the St. Francis Reservoir in 1926 provided critical storage for stabilizing water and power during the 1927 bombings.

The Owens Valley aggression ended after arrest of the primary instigators for embezzlement. The City obtained financial statements suggesting illegal diversion of bank funds in the Owens Valley [76], which were presented to authorities in August 1927 and led to bank closure and arrest. This evidence was uncovered by an LADWP supporter in the Owens Valley and the LADWP Council following suspicions that bank owners were financing the bombings [78].

WWI, WWII, and Cold War

Special security and protection measures were implemented by the Water System and civil defense agencies during WWI, WWII, and the cold war era. Armed guards from the U.S. Military, City police, and LADWP security patrolled LADWP facilities. Figure 36 shows Military guard patrols. Guard dogs also helped patrol and secure facilities. Other measures were taken at critical facilities such as security lighting and enclosures.

a.   b.

Figure 36. WWII armed military guards protecting; a) the LAA, and b) City facilities.

Security protections continued in the 1950's and 1960's for the cold war. Military guards were no longer used, but patrols continued for critical facilities and the LADWP worked with the California Office of Civil Defense and the American Red Cross to remain prepared for possible attacks. Mock air raids and other attack exercises were periodically practiced to keep employees trained in emergency response and recovery. Aqueduct personnel regularly attended domestic terrorism courses at the California Specialized Training Institute in San Louis Obispo, California. Figure 37 shows a Civil Defense truck checking for radiation in a reservoir. These types of trucks were equipped for response to many different types of emergencies.

Figure 37. Cold war radiation check at LADWP reservoir (1955).

EMERGENCY PREPAREDNESS AND RECOVERY

Emergency preparedness encompasses a large program of being prepared for disasters through training, planning, capital improvement, system maintenance and rehabilitation, equipment and repair materials inventory, emergency response and recovery plans, system management, risk management strategies, risk assessment, hazard assessment and evaluation, making cooperative agreements with other water agencies, etc., which are incorporated into the overall LADWP emergency preparedness and contribute to the historical Water System multihazard improvements. In essence, emergency preparedness encompasses much of what has been discussed in this report and more with a primary goal of allowing the post-disaster recovery to begin, proceed, and be completed as soon and as efficient as possible, and to reduce the possibility of error; however descriptions of these aspects are beyond the scope of this report. Los Angeles Water System emergency preparedness incorporates cooperative effort with other Los Angeles City departments and emergency response organizations throughout the City and County of Los Angeles, the State of California, and the United States; the LADWP maintains an emergency preparedness office, and the City has a similar Department to plan and coordinate much of these efforts with the other City Departments and government agencies. This section only deals with some emergency preparedness aspects the Water System has instituted to respond to a disaster and aid system recovery.

Emergency response plans in the Los Angeles Water System are generally developed in preparation for an earthquake because historically earthquakes have created the greatest disasters, but the plans are applicable to other disasters. Because all available personnel are assigned duties in order to inspect all critical parts of the system as soon as possible after major earthquakes, most other emergencies can be handled using a small part of the emergency response plans. Some of these activities were initiated following the 1971 San Fernando Earthquake.

The first line of response is the operating business units whose operators are pre-assigned specific facilities to inspect. These personnel typically inspect the same facilities they operate on a daily basis. They are equipped with cellular phones and Department radios that are installed in vehicles. The LADWP maintains their own microwave system and radio frequencies for communication. Field personnel communicate through normal lines, which include their supervisors and 24-hour control centers. Staffing at control centers is supplemented as soon as possible.

Engineering personnel are assigned emergency response duties related to their normal duties and expertise. Damage Assessment Teams (DATs) are multidisciplinary teams assigned to inspect structures and trunk lines in specific parts of the city. They report through a team leader whose responsibility is to support a district superintendent in making evaluations of facilities, designing repairs, and estimating repair costs to be used in obtaining government financial assistance. Reservoir Inspection Teams (RIT) are tasked specifically with inspections of the Water Systems dams and reservoirs. The RIT respond to problems reported by operators and also make their own detailed inspections of dams to ensure their continuing safe and reliable performance.

During major emergencies, Water System Management staffs a command center, called the Water Emergency Control Center, which is designed to gather damage and status reports in one location and expedite recovery planning, set priorities, and pass requests for assistance between various work units. LADWP also helps staff the City's Emergency Operations Center to coordinate with other agencies and City Departments.

To supplement the emergency response and recovery plan, the Water System maintains an inventory of pipe, fittings, valves, repair clamps, construction materials, etc., construction equipment, and trained personnel that are used to repair the system from regular operational problems such as a normal pipe break, and after a disaster such as an earthquake, which may cause many pipe leaks and other system wide damages. The Water System operation and maintenance and emergency response are performed at seven separate operation and maintenance field headquarters, which can operate independently or supplement each other. Within the City the distribution system is broken into five districts, each having their own maintenance yards. The LAA is broken into two districts, each having their own maintenance yards. Having several district yards located throughout the large Los Angeles metropolitan area has shown great advantages in post-disaster recovery in that disasters have not affected all districts at the same time, which allows the unaffected districts to supplement the supplies and resources within the affected districts. This type of district support proved very effective and beneficial following the 1971 San Fernando and 1994 Northridge Earthquakes. In both earthquakes the LAA Southern District Mojave yard was also able to provide support to distribution repairs in addition to making the LAA repairs, showing how all seven maintenance yards can provide support as necessary, depending on the disaster.

An example of emergency preparedness includes the Elizabeth Tunnel Shaft at the northerly end of the LAA Elizabeth Tunnel. The Elizabeth Tunnel was excavated through the San Andreas Fault zone and all LAA water passes through the tunnel. The inclined shaft permits direct access by equipment to repair the tunnel if damaged by an earthquake. A small diameter shaft has been constructed at the approximate tunnel midpoint for the installation of electric cables, communication cables, etc. to the tunnel below. Spare aqueduct pipe is stored at various locations.

DISCUSSION

The numerous examples of multihazard mitigation for the Los Angeles Water System reveal how mitigation for one hazard impacts other hazards. Figure 38 presents a hazard mitigation matrix that summarizes the interrelationship between performing mitigation for the hazard listed in the left column and impacts to the hazard mitigations in the top row. Figure 38 is only based on Los Angeles Water System experiences identified by the authors, some of which are not described in this report, and is not considered comprehensive for all mitigation interactions that have affected the Water System. Figure 38 is presented in an upper diagonal matrix by consideration that the hazard mitigations inversely correlate with each other as a result of: (1) multiple hazards being considered when implementing mitigations and (2) mitigation efforts providing inherent byproducts that unintentionally impact other hazards, either positively or negatively, that are recognized after implementation.

Sometimes the mitigation impact is positive in that the primary mitigation solution also provides mitigation for other hazards; for example, mitigation for drought has led the LADWP to securing multiple water supplies and storage, which has aided in developing a resilient system for recovering from earthquake damage. Sometimes the impact is negative and inhibits mitigation for other hazards; for example some water quality improvements have reduced available storage reservoirs and inhibit the systems capability to recover rapidly from strong earthquake damage, requiring additional seismic improvements. Still in other cases the effect of multiple hazards has been evaluated in determining how to change and improve the system; for example several dams have been removed from service through diligent consideration of earthquake, dam safety, and reservoir water quality concerns. The complications and difficulties in understanding multiple hazard interrelationships are revealed in Figure 38 and as a consequence indicate the need for managing multihazard mitigations.

Mitigation for Hazard:	Impacts Mitigation for (positive, negative, or in conjunction):														
	Earthquake	Pipe Deterioration	Storm Water/Flood	Debris Flow	Drought	Fire	Landslide	Volcanism	Soil Settlement	Cold Weather	Wind/Seiche	Water Quality	Hazardous Material	Dam & Res. Safety	Homeland Security
Earthquake	*	*			*	*	*	*	*	*	*	*	*	*	*
Pipe Deterioration		*					*		*	*		*			
Storm Water/Flood			*	*								*	*	*	
Debris Flow				*		*	*	*				*	*	*	
Drought					*							*		*	*
Fire						*						*	*		*
Landslide							*				*	*	*	*	
Volcanism								*				*	*	*	
Soil Settlement									*				*		
Cold Weather										*				*	
Wind/Seiche											*	*	*	*	
Water Quality												*	*	*	*
Hazardous Material													*		
Dam & Res. Safety														*	*
Homeland Security															*

Figure 38. Matrix showing relationship between different hazard mitigations from LADWP Water System Historical experiences.

In addition, there are numerous examples of multiple hazards combining to create disasters. For example, the St. Francis Dam disaster resulted from a combination of landslide, dam safety, and flood following the dam failure. Debris inundation of the Upper and Middle Debris Basins resulted from a combination of earthquake induced landslides and El Niño storm induced floods and debris flows. The combined interactions of multiple hazards are important to the engineering design and system operations, and much is to be learned from case studies similar to those presented herein.

Figure 39 presents a compounding hazard interaction matrix summarizing how different hazards may combine to increase a potential disaster. Figure 39 was developed based on Los Angeles Water System experiences as identified by the authors and obtained through design concepts, studies of actual disasters, and observation of occurrences where disasters were preempted, not all of which are described herein. This matrix is not considered comprehensive for all compounding hazards affecting the Los Angeles Water System. Figure 39 should be read by first considering a primary hazard in the left column and how effects from this hazard may be compounded by those listed at the top. For example, an earthquake may cause pipe ruptures and fires, and fire fighting may be inhibited due to a reduced water supply directly resulting from the pipe breaks. Thus, Figure 39 may indicate a hazard cause and effect, such as the earthquake causing pipe breaks, or a direct hazard combination such as pipe breaks/lack of water and fire.

Hazard:	Combined with this Hazard to Create Greater Disaster:														
	Earthquake	Pipe Deterioration	Storm Water/Flood	Debris Flow	Drought	Fire	Landslide	Volcanism	Soil Settlement	Cold Weather	Wind/Seiche	Water Quality	Hazardous Material	Dam & Res. Safety	Homeland Security
Earthquake	X	+	+	+		+	+	+	+	+	+	+	+	+	
Pipe Deterioration		X		+		+	+	+	+	+		+	+	+	+
Storm Water/Flood			X	+		+	+				+	+	+	+	
Debris Flow				X		+						+	+	+	
Drought					X	+									
Fire			+	+		X	+					+	+	+	
Landslide		+		+			X					+	+	+	
Volcanism	+	+		+			+	X				+	+	+	
Soil Settlement		+							X				+		
Cold Weather		+								X				+	
Wind/Seiche		+					+				X	+		+	
Water Quality												X			+
Hazardous Material				+								+	X		+
Dam & Res. Safety			+										+	X	+
Homeland Security			+			+						+	+	+	X

Figure 39. Matrix showing how multiple hazards may combine to create greater disaster from Los Angeles Water System historical experiences.

As seen in Figure 39, more hazards combine with earthquake to develop compounding disaster effects than any other primary hazard. In addition, dam and reservoir safety and water quality are the two most common hazards that may combine with other hazards. Figure 38 shows that mitigations for earthquake, dam and reservoir safety, and water quality, are the three greatest hazard mitigations that impact other mitigations. These tabulations help clarify why earthquake, water quality, and dam and reservoir safety are the most common hazards the LADWP Water System deals with.

The competition and prioritization of resources and the determination if mitigation is appropriate to implement in some cases are two very important issues to consider in hazard mitigation. This is understood through examples from post-1971 earthquake mitigation as follows:

The primary lesson learned from the 1994 tank performance vs. the post 1971 SVA experience is how competing priorities can affect the seismic improvement program and how implementation of programs can fade over long periods of time, especially for the lower priority issues. For example, there were no significant damages to dams, pumping stations, or chlorination stations because these post-1971 improvements had very high priorities for completing. However, as in the tank example, some tank and sump roofs and inlet-outlet lines were improved after 1971, but several were not due to conflicting priorities in managing necessary water capital improvement projects, primarily for water quality. There are similar examples for pipe performances in liquefaction areas, equipment anchorages, and flexibility of water conveyance pipelines, where not all lessons were adequately implemented (e.g., equipment anchorages) or they were found infeasible (e.g., relocating pipes out of liquefaction zones) and preferable to sustain damages and then proceed with repairs following the earthquake.

CONCLUSION

The LADWP has a long and illustrious history of Water System multihazard mitigations and improvements. Many examples from fifteen different hazards affecting the Los Angeles Water System have been described herein; however, there are many other examples that are not mentioned. The LADWP has made historical impacts in the water and engineering community including improvements in dam and reservoir safety, developing the soil compaction methods that are most commonly used worldwide, developing and implementing seismic evaluation and construction methodologies for embankment dams, methods for the design, construction, and rehabilitation of pipelines, drought mitigation, water quality improvements, and so on. In addition, the LADWP has a strong emergency response program that helps mitigate effects of all potential hazards.

The LADWP has learned much from actual disaster experiences and has provided significant efforts in documenting, preparing case studies, researching, and providing others with all available information in order to learn and improve the Los Angeles and other water systems from these experiences. The hazard improvements established throughout LADWP's history have proven beneficial in withstanding disaster effects, including the 1971 San Fernando and 1994 Northridge Earthquakes, by allowing the system to continue operating after suffering damage. Performance during these two earthquakes and other disasters clearly shows the disaster resiliency developed within the City of Los Angeles' Water System.

Case studies of the Los Angeles Water System show that mitigation for one type of hazard many times also help to mitigate other hazards, intentionally or unintentionally; several examples of this are provided. On the other hand, the studies also showed that mitigation for one hazard can have negative impacts against other hazard mitigations. In addition, there are many cases where several different hazards have combined to increase the disaster effects. Tables summarizing these conditions from the Los Angeles Water System experiences were developed. These types of interactions clarify the need for multihazard management.

A review of the Los Angeles history shows that all damage cannot be eliminated. It is best to be prepared for potential disasters, have the capacity to bounce back from the damage, and be versatile in the ability to supply water. A combination of strengthening to help keep damage from occurring, providing redundancy in supplies and supply lines to allow water to flow after damage has occurred, incorporating isolation capabilities so that damaged portions of the distribution system do not drain all the resources, and an adequate well-trained staff to respond to the disaster is needed.

The Los Angeles Water System is currently undergoing significant system-wide changes to improve the quality of the water supplied to customers. In responding to this challenge the LADWP has incorporated effects of other hazards that include a seismic improvement program to understand post-earthquake system performance and re-evaluate the seismic stability of LADWP dams, general dam safety, flooding, pipe rehabilitation and replacement, homeland security, and others. The LADWP's goal is to continuously improve the resiliency and performance capabilities of the Los Angeles Water System.

ACKNOWLEDGEMENTS

This document was originally prepared for the Third US-Japan Workshop on Water System Seismic Practices, Los Angeles CA., August 6 to 8, 2003 for the purpose of documenting the Los Angeles Water System history of seismic improvements [79]. This current report expands upon [79] to include the multihazard mitigation historical perspective of the Los Angeles Water System. This report was originally presented during the pre-conference workshop in conjunction with the 6th US Conference on Lifeline Earthquake Engineering (TCLEE 2003) on August 10, 2003, Long Beach, CA.

The authors would like to thank and acknowledge the many LADWP employees and others who have aided in developing multihazard improvements into the Los Angeles Water System over the past 100 years; Catherine Mulholland for sharing the notes on her presentation given at the 2004 Earthquake Engineering Research Institute Annual Meeting Banquet; George Brodt for review of the seismic portions of this manuscript; Cliff Plumb and Larry Jackson for providing information for landslides; Cliff Plumb for providing information on volcanism; Craig Bagaus for providing information on the 2003 debris flow; Phil Lahr for providing emergency response and dam safety information; Phil Clark for providing tank inlet/outlet flexible coupling information; Ali Karimi for providing information for water quality and hazardous materials; Melinda Rho for providing information on water quality standards; James Campbell and Charlotte Rodrigues for providing information for spillway repairs at Grant Lake and Long Valley; Charles Parkes for providing information on Type IIA cement; Douglas Sunshine for providing homeland security information; V. Miller for reviewing the Homeland Security section; James Yannotta for reviewing the Homeland Security and Hazardous Materials sections; Ernie Havalina and Luis Castillo for providing information on cross-connections; Gary Mackey, Steve Torres, and Leland Gong for providing information on hazardous materials; Victor Murillo, Kien Huang, and Roger Callo for drafting figures. The authors provide a special acknowledgement to Federico Waisman, John Sturman, Dennis Ostrom, Amar Chaker, and Craig Taylor for their reviews of the manuscript and valuable and insightful comments. All photographs are from the LADWP files unless otherwise noted.

REFERENCES

[1] **Mulholland, C.**, 2000, "William Mulholland and the Rise of Los Angeles," *University of California Press*, Berkeley.

[2] **Rogers, J. D.**, 1995, "A Man, A Dam and A Disaster: Mulholland and the St. Francis Dam," in "The Saint Francis Dam Disaster Revisited," Doyce B. Nunis, Jr., Ed., *Historical Society of Southern California and Ventura County Museum of History and Art*, Los Angeles, Ventura.

[3] **Mulholland, C.**, 2004, "William Mulholland; An Engineer in Earthquake Country," presentation manuscript at the 2004 Earthquake Engineering Research Institute Annual Meeting Dinner Banquet, Feb. 6, 2004.

[4] **Outland, C. F.**, 2002, "Man-Made Disaster the Story of St. Francis Dam," 2nd printing of the revised edition, *The Ventura County Museum of History and Art*, Ventura, California.

[5] **Van Norman, H. A.**, 1934, "Bouquet Canyon Reservoir and Dams Large Earth-fill Structure Provides Needed Storage for Domestic Supply of Los Angeles," *Civil Engineering*, ASCE, Aug. 1934.

[6]   **International Conference of Building Officials (ICBO)**, 1964, "Uniform Building Code," 1964 Ed., Whittier, California, May.

[7]   **Bardet, J. P. and C. A. Davis**, 1996, "Engineering Observations on Ground Motion at the Van Norman Complex after the Northridge Earthquake," *Bulletin of Seismological Society of America*, Special Northridge Issue, Vol. 86, No. 1B, pp. S333-S349.

[8]   **Proctor, R. R.**, 1933, "Fundamental Principals of Soil Compaction," *Engineering News Record*, Aug. 31, Sept. 7, Sept. 21, Sept. 28.

[9]   **Nelson, S. B. S.**, 1941, "Heavy Construction of Stormdrains to Protect San Fernando Reservoirs," *Southwest Builder and Contractor*, Feb. 7, 1941, pp. 20-23.

[10] **Los Angeles Department of Water and Power**, 1966, "Lower San Fernando Dam Slope Stability Analysis," *Water Engineering Design Division* Report AX 215-31.

[11] **Seed, H. B, K. L. Lee, I. M. Idriss, and F. I. Makdisi**, 1973, "Analysis of the Slides in the San Fernando Dams During the Earthquake of February 9, 1971," *Report No. UCB/EERC 73-2* University of California, Berkeley, California.

[12] **Wald, D. J., and T. H. Heaton**, 1994, "A Dislocation Model of the 1994 Northridge, California, Earthquake Determined from Strong Ground Motions," *United States Department of the Interior, U.S. Geological Survey*, Open-File Report 94-278.

[13] **Weber, F. H.**, 1975, "Surface Effects and Related Geology of the San Fernando Earthquake in the Sylmar Area," Chapter 6 in San Fernando, California, Earthquake of 9 February 1971, Oakeshott, G. B., Ed., *California Division of Mines and Geology, Bulletin 196*, Sacramento, CA.

[14] **Scott, R. F.**, 1973, "The Calculation of Horizontal Accelerations from Seismoscope Records," *Bulletin of the Seismological Society of America*, Vol. 63, No. 5, pp. 1637-1661.

[15] **Cortright, C. J.**, 1975, "Effects of the San Fernando Earthquake on the Van Norman Reservoir Complex," Chapter 29 in San Fernando, California, Earthquake of 9 February 1971, Oakeshott, G. B., Ed., *California Division of Mines and Geology, Bulletin 196*, Sacramento, CA.

[16] **Subcommittee on Water and Sewage Systems**, 1973, "Earthquake Damage to Water and Sewerage Facilities," *San Fernando, California, Earthquake of February 9, 1971*, N. A. Benfer and J. L. Coffman eds., U. S. Dept. of Commerce, NOAA Spec. Rpt., Vol. II, pp. 75-193.

[17] **O'Rourke, T. D. and M. Hamada, Eds.**, 1992, "Case Studies of Liquefaction and Lifeline Performance during Past Earthquakes," *NCEER-92-0002*, Vol. 2, National Center for Earthquake Engineering Research, Buffalo, NY.

[18] **Dames and Moore**, 1985, "Evaluation of Earthquake-Induced Deformations of Pleasant Valley Dam," Report prepared for the City of Los Angeles Department of Water and Power, Dames and Moore Job No. 00138-015-02.

[19] **Roth, W. H., Bureau, G., and Brodt, G.**, 1991, *Pleasant Valley Dam" An Approach to Quantifying the Effect of Foundation Liquefaction*, Int. Conference, Commission of Large Dams, Vienna, Austria.

[20] **Dames and Moore**, 1991, "Stability Evaluation South Haiwee Dam Inyo County, California," Report prepared for the City of Los Angeles Department of Water and Power, Dames and Moore Job No. 00138-071-015.

[21] **Los Angeles Department of Water and Power**, 1974, "Report of Water System Vulnerability to Earthquakes," L. Lund and R. Triay, *Water Engineering Design Division Report AX 275-28*.

[22] **Miedema, H. J., and J. B. Olson**, 1974,"Repair of Seismic Damage to Above Ground Pipelines," *Transportation Engineering Journal, ASCE*, 100, TE3, pp. 733-742.

[23] **Davis, C. A.**, 1999, "Performance of a Large Diameter Trunk Line During Two Near-Field Earthquakes, *Proc. 5th U.S. Conf. on Lifeline Earthquake Engr*, ASCE, Seattle, Aug., pp. 741-750.

[24] **Davis, C. A.**, 2001, "Retrofit of a Large Diameter Trunk Line Case Study of Seismic Performance," *Proc. of 2nd Japan-US Workshop on Seismic Measures for Water Supply*, AWWARF/JWWA, Tokyo, Japan, Aug. 6-9, American Waterworks Association Research Foundation Project 2786, Session 2.

[25] **Davis, C. A. and S. R. Cole**, 1999, "Seismic Performance of the Second Los Angeles Aqueduct at Terminal Hill," *Proc. 5th U.S. Conf. on Lifeline Earthquake Engr*, ASCE, Seattle, Aug., 452-461.

[26] **Toprak, S.**, 1998, "Earthquake Effects on Buried Lifeline Systems," Dissertation presented to Cornell University in partial fulfillment for the degree of Doctor of Philosophy.

[27] **Lund, L., and E. Matsuda**, 1994, "Lifelines," *Northridge Earthquake, January 17, 1994, Preliminary Reconnaissance Report*, Chapt. 6, Earthquake Engineering Research Institute, pp. 67-77.

[28] **O'Rourke, T. D. and M. J. O'Rourke**, 1995, "Pipeline Response to Permanent Ground Deformation: A Benchmark Case," *Proc. 4th U.S. Conf. on Lifeline Earthquake Engineering,* ASCE, San Francisco, Aug., pp. 288-295.

[29] **Davis, C. A. and J. P. Bardet**, 1995, "Seismic Performance of Van Norman Water Lifelines," *Proc. 4th U.S. Conf. on Lifeline Earthquake Engineering,* ASCE, San Francisco, Aug., pp. 652-659.

[30] **Davis, C. A., and J. P. Bardet**, 1996, "Performance of Two Reservoirs During 1994 Northridge Earthquake," *J. Geotech. Engrg. Div.,* ASCE, Vol. 122, No. 8, pp. 613-622

[31] **Brown, K., P. Rugar, C. Davis, and T. Rulla**, 1995, "Seismic Performance of Los Angeles Water Tanks," *Proc. 4th U.S. Conf. on Lifeline Earthquake Engr,* ASCE, San Francisco, Aug., pp. 668-675.

[32] **Bardet, J. P. and C. A. Davis**, 1995, "Lower San Fernando Corrugated Metal Pipe Failure," *Proc. 4th U.S. Conf. on Lifeline Earthquake Engineering,* ASCE, San Francisco, Aug., pp. 644-651.

[33] **Davis, C. A. and J. P. Bardet**, 1998, "Seismic Analysis of Large Diameter Flexible Underground Pipes," *J. Geotech. and GeoEnv. Engrg. Div.,* ASCE, Vol. 124, No. 10, pp. 1005-1015.

[34] **Davis, C. A.**, 2000, "Study of Near-Source Earthquake Effects on Flexible Buried Pipes," *Dissertation presented to the University of Southern California in partial fulfillment for the degree of Doctor of Philosophy.*

[35] **Youd, T. L.**, 1971, "Landsliding in the Vicinity of the Van Norman Lakes," in The San Fernando, California, Earthquake of February 9, 1971, U. S. Geological Survey *Prof. Paper 733,* pp. 105-109.

[36] **Los Angeles Department of Water and Power**, 1997, "Response of the High Speed and Bypass Channels to the 1994 Northridge Earthquake and Recommended Repairs," *Water Supply Division Report AX 215-47.*

[37] **Davis, C. A., J. P. Bardet, and J. Hu**, 2002, "Effects of Ground Movements on Concrete Channels" *Proc. 8th US-Japan Workshop on Eq Resistant Des. of Lifeline Fac. and Countermeasures for Soil Liquefaction,* Tokyo, T. D. O'Rourke, J. P. Bardet, and M. Hamada, Eds., Tech. Rep. MCEER-03-0003, pp 111-122.

[38] **Archuleta, R. J., G. Mullendore, and L. F. Bonilla**, 1998, "Separating the Variability of Ground Motion over Small Distances," Proceedings from *The Effects of Surface Geologyon Seismic Motion,* Balkema, Rotterdam.

[39] **Trifunac, M. D., M. I. Todorovska, and V. W. Lee**, 1998, "The Rinaldi Strong Ground Motion Accelerogram of the Northridge, California Earthquake of 17 January 1994," *Earthquake Spectra,* EERI, Vol. 14, No. 1, pp. 225-239.

[40] **Cultrera, G., D. M. Boore, W. B. Joyner, C. M. Dietel**, 1999, "Nonlinear Soil Response in the Vicinity of the Van Norman Complex Following the 1994 Northridge, California, Earthquake," *Bulletin of the Seismological Society of America,* Vol. 89, No. 5, pp. 1214-1231.

[41] **Sano, Y., T. D. O'Rourke, and M. Hamada**, 2002, "Permanent Ground Deformation due to Northridge Earthquake in the Vicinity of the Van Norman Complex, *Proc. 7th US-Japan Workshop on Earthquake Resistant Des. of Lifeline Fac. and Countermeasures for Soil Liquefaction,* Seattle, T. D. O'Rourke, J. P. Bardet, and M. Hamada, Eds., Tech. Rep. MCEER-99-0019.

[42] **Bardet, J. P., J. Hu, T. Tobita, and N. Mace**, 2002, "Large-Scale Modeling of Liquefaction-induced Ground Deformation," Earthquake Spectra, Vol. 18, No. 1, pp. 19-46.

[43] **Toprak, S., O'Rourke, T. D., and Tutuncu, I.**, 1999, "GIS Characterization of Spatially Distributed Lifeline Damage," Proceedings, 5th US Conference on Lifeline Earthquake Engr, Seattle, WA,

[44] **Bardet, J. P., J. Hu, T. Tobita, and N. Mace**, 1999, "Database of Case Histories on liquefaction-induced Ground Deformation," A report to PEER/PG&E, Task 4 – Phase 2, October, Civil Engineering Department, University of Southern California, Los Angeles, CA.

[45] **American Lifelines Alliance**, 2001, "Seismic Fragility Formulations for Water Systems," John Eidinger Principal Investigator, www.americanlifelinesalliance.org.

[46] **O'Rourke, T. D., Y. Wang, P. Shi, and S. Jones**, 2004, "Seismic Wave Effects on Water Trunk and Transmission Lines," 11th International Conference on Soil Dynamics & Earthquake Engineering, Berkeley, CA, Jan. 7~9, Submitted.

[47] **Tutuncu, I., T. D. O'Rourke, J. A. Mason, and T. K. Bond**, 2002 "Seismic Rehabilitation of Water Trunk Lines with Fiber Reinforced Composite Wraps", *Proceedings, 7th National Conference on Earthquake Engineering,* Boston, MA, July, EERI, Oakland, CA..

[48] **O'Rourke, T. D.**, 2003, Professor, Personal communication, School of Civil and Environmental Engineering, Cornell University, Ithaca, N.Y.

[49] **URS Greiner Woodward Clyde**, 2000, "Review of Seismic Stability of Seventeen Water Service Organization Dams," Prepared for the Los Angeles Department of Water and Power.

[50] **URS**, 2002, "Seismic Stability Evaluation of Stone Canyon Dam," Report prepared for the City of Los Angeles Department of Water and Power, URS Project No. 59-00112105.01, Water Services Organization Report No. AX 213-38.

[51] **Roth, W. H., E. M. Dawson, C. A. Davis, and C. C. Plumb**, 2004, "Evaluating the Seismic Performance of Stone Canyon Dam with 2-D and 3-D Analyses," *13th World Conference on Earthquake Engineering*, August 1-6, Vancouver, BC, Canada, in press.

[52] **Davis, C. A.**, 2003, "Lateral Seismic Pressures for Design of Rigid Underground Structures, *Proc. 6th U.S. Conf. on Lifeline Earthquake Engineering*, ASCE, Long Beach, Aug., 10-13, pp. 1001-1010

[53] **Keilis-Borok, V. I., and A. A. Soloviev**, 2003, "Nonlinear Dynamics of the Lithosphere and Earthquake Prediction," Eds., *Springer*, Berlin.

[54] **Los Angeles Department of Water and Power**, 1996, "Trunk Line Condition Assessment Program – First Phase," *Water Supply Division*.

[55] **Los Angeles Department of Water and Power**, 1998, "Trunk Line Condition Assessment Program – Second Phase," *Water Supply Division*.

[56] **Lund, L.**, 2003, "Pipeline Seismic Mitigation using Trenchless Technology," *Proc. 6th U.S. Conf. on Lifeline Earthquake Engineering, Monograph 25*, ASCE, Long Beach, Aug., 10-13, pp. 736-743.

[57] **Los Angeles Department of Water and Power**, 1997, "Response of the Middle Debris Basin to the 1994 Northridge Earthquake and Recommended Repairs, *Water Supply Division Report No. AX 215-46*.

[58] **Harp, E. L., and R. W. Jibson**, 1996, "landslides Triggered by the 1994 Northridge, California, Earthquake," *Bull. Seism. Soc. Am.* – Special Northridge Issue 86, No. 1B, pp. S319-S332.

[59] **Los Angeles Department of Water and Power**, 2000, "Urban Water management Plan," Fiscal Year 2002-2003 update, *Water Resources Business Unit*.

[60] **Los Angeles Department of Water and Power**, 1995, "Intersystem Pumping Location Maps," Water Operating Division Graphics, Maps, and Records Group.

[61] **Los Angeles Department of Water and Power**, 1956, "Repair of Earth Slide and Construction of Compacted Earth Slope and Bottom Blankets for Upper Stone Canyon Reservoir," *Field Engineering Division Report, AX-213-32*.

[62] **Los Angeles Department of Water and Power**, 1998, "Contingency Plan for Long Valley Volcanic Eruptions" Water Services Organization Report No. AX 522-1.

[63] **Phillips, R. V., and S. M. Marynick**, 1972, "Brittle Fracture of Steel Pipeline Analyzed", *Civil Engineering*, American Society of Civil Engineers, New York, NY, July 1972.

[64] **Los Angeles Department of Water and Power**, 1971, "The Second Los Angeles Aqueduct," *Los Angles Department of Water and Power*, Los Angeles.

[65] **Georgeson D. L. and A. A. Karimi**, 1988, "Water Quality Improvements with the use of Ozone at the Los Angeles Water Treatment Plant," *Ozone Science & Engineering*. 10(3): 255-276.

[66] **Karimi, A. A., J. A. Redman, W. H. Glaze, and G. F. Stolarik**, 1997, "Evaluating an AOP for TCE and PCE Removal," *Journal of American Water Works Association*, Vol. 89, August.

[67] **Karimi, A. A., S. Adham, S. Tu**, 2002, "Evaluation of Membrane Filtration for Los Angeles' Open Reservoirs," *Journal of American Water Works Association*, Vol.94, No. 12, December.

[68] **Karimi, A. A., D. Wetstein, R. Trussell, S. Hirai and G. Stolarik**, 2002, "A Chloramine Conversion Evaluation of the Nations most Complex Water System," *Proceedings of the AWWA Annual Conference and Exposition*. New Orleans, Louisiana, June 16-20.

[69] **Cloran, W., W. Tibbetts, and E. Alterton**, 1948, *Foundation for Cross-Connection Control and Hydraulic Research, Paper No. 5*, University of Southern California.

[70] **Reynolds, K., and E. Springer**, 1959, "Definitions and Specifications of Double Check Valve Assemblies and Reduced Pressure Principles Backflow Prevention Devices," *Foundation for Cross-Connection Control and Hydraulic Research, USCEC Report 48-101*, University of Southern California, under LADWP Contract 10609.

[71] **Foundation for Cross-Connection Control and Hydraulic Research**, 1960, "Manual of Cross Connection Control," University of Southern California.

[72] **Hamilton, D. H, and R. L. Meehan**, 1971, "Ground Rupture in the Baldwin Hills," *Science*, Vol. 172, pp. 333-344.

[73] **Grant, U. S.**, 1964, "Geologic Report on the Destruction of the Baldwin Hills Reservoir," *Consultant* report prepared for Roger Arnebergh, City Attorney of Los Angeles.

[74] **Department of Water Resources**, 1979, "Dams within the Jurisdiction of the Sate of California," *State of California*, The Resource Agency, Bulletin 17-79.

[75] **Davis, M. L.**, 1993, "Rivers in the Desert," *HarperCollins*, New York.

[76] **Kahrl, W. L.**, 1982, "Water and Power: The Conflict Over Los Angeles' Water Supply in the Owens Valley," *University of California Press*, Berkeley.

[77] **Nadeau, R.**, 1950, "The Water Seekers," *Doubleday*, Garden City, New York.

[78] **Hoffman, A.**, 1981, "Vision or Villainy Origins of the Owens Valley – Los Angeles Water Controversy," *Texas A&M University Press*, College Station.

[79] **Davis, C. and L. Lund**, 2003, "History of Los Angeles Water System Improvements," *Proc. of 3ⁿᵈ US-Japan Workshop on Water System Seismic Practices*, AWWARF/JWWA, Los Angeles, Aug. 6-8, American Waterworks Association Research Foundation Project, Session 4.

APPENDICES

APPENDIX A

SHORT BIOGRAPHIES OF CONTRIBUTING AUTHORS

Beverley Adams, Ph.D., is Remote Sensing Group Leader at ImageCat, Inc. Dr. Adams specializes in the application of advanced technologies for post-disaster damage assessment, disaster management and loss estimation, and was recently invited to brief the White House Homeland Security Council Members about her research. She has ten years experience in the field of remote sensing, and has lead projects involving multispectral airborne, satellite and UAV data, lidar imagery, stereoscopic photography and SAR data. Dr Adams has responded to several international earthquakes and coordinated windstorm field reconnaissance activities in the aftermath of Hurricanes Charley and Ivan. Dr. Adams acted as a consultant to the British National Space Center while completing her PhD at University College London. During this time, she also served as Secretary to the International Society of Photogrammetry and Remote Sensing Commission II. Dr. Adams has a keen interest in public outreach, education and communication. She is a seasoned public presenter, recently serving as a plenary speaker for the EERI Bam earthquake briefings and chairing the 2nd International Workshop on Remote Sensing for Disaster Response. Dr. Adams currently serves on the NSF CMS review panel, and is a reviewer for the journals Earthquake Spectra and Photogram metric Engineering and Remote Sensing.

Jose Carlos Borrero, Ph.D., currently serves as a research associate professor in the Department of Civil and Environmental Engineering at the University of Southern California. In 1993, he received a bachelor's degree from the University of Florida in the School of Journalism and then a B.S., M.S., and Ph.D. from the University of Southern California in the Civil and Environmental Engineering. His research interests include tsunamis; their generation, propagation and coastal effects. Dr. Borrero has participated in numerous post tsunami filed surveys to see first hand their effects. His current project is the creation tsunami inundation and hazard maps for the state of California.

Amar Chaker, Ph.D. is currently Director of Engineering Applications at the Civil Engineering Research Foundation, after serving as Director of ASCE's Transportation and Development Institute, and staff contact for several Divisions and Councils within ASCE's Technical Activities, including the Technical Council on Lifeline Earthquake Engineering and the Council on Disaster Reduction. He has held faculty positions at the University of Illinois at Urbana-Champaign and Drexel University. He has served as Professor and Director of the Civil Engineering Institute of the University of Science and Technology of Algiers, Algeria. As Technical Director of the Organization for Technical Control of Construction, (CTC) in Algeria he co-chaired the Committee for the Algerian Earthquake-Resistant Design Code, and participated in the development of the 1981, 1983 and 1988 editions. He also participated in post-earthquake investigations and in a seismic hazard evaluation and urban microzonation study for the region of El Asnam, Algeria. He holds a Ph.D. degree in Civil Engineering from the University of Illinois at Urbana-Champaign and a degree of 'Ingénieur Civil' from 'Ecole Nationale des Ponts et Chaussées', Paris, France.

Sungbin Cho is a specialist in regional economic models and transportation systems analysis. He received his BA in engineering (1988) and MA in planning (1990) from Seoul National University. At present, he is a PhD candidate in Urban Planning at the University of Southern California's School of Public Policy and Development in Los Angeles, and a Transportation Systems Analyst with ImageCat, Inc. in Long Beach. His dissertation research focuses on development of a computable general equilibrium model of the urban economy. Since 1990, Mr. Cho has been a consultant for numerous transportation planning projects with the South Korean government and the Seoul metropolitan government, including improvement of the Korean peninsular freeway system, and development of the Seoul subway system. He arrived at USC in 1996, working with faculty collaborators to develop and apply an integrated, urban economic - transportation analysis model of the Los Angeles region, the Southern California Planning Model (SCPM, versions 2.0 and 2.5). During the same period, he contributed to the development of the REDARS (Risks from Earthquake Damaged to Roadway Systems) software, working with researchers at USC and ImageCat, Inc.

Craig Davis, Ph.D., PE, GE is the Geotechnical Engineering Manager for the Los Angeles Department of Water and Power, Water System and is responsible for managing geotechnical engineering, geology, and soils and material testing for LADWP projects and overseeing dam and reservoir safety including monitoring, surveillance, stability evaluations, and design. He is a California licensed Civil and Geotechnical Engineer and received a B.S. in Civil Engineering from the California Polytechnic State University in San Louis Obispo, CA, an M.S. in Civil Engineering with emphasis in structural earthquake engineering from the University of Southern California in 1991, and a Ph.D. in Civil Engineering with emphasis in geotechnical earthquake engineering from the University of Southern California in 2000.

He has worked for the LADWP since 1987 where he has managed several multimillion dollar projects and implemented unique and innovative designs including the use of stone columns for controlling differential settlement and Teflon bearing surface base isolation to mitigate effects of liquefaction-induced permanent ground deformation. He is actively involved in earthquake, geotechnical, and lifeline engineering research, developing a number of case studies from the 1971 San Fernando and 1994 Northridge earthquakes and performing evaluations on the seismic performance of dams and reservoirs, pipelines and underground structures, and the effects of ground deformations on water system facilities. Dr. Davis has published over 40 papers in technical journals and conference proceedings and has co-organized national and international workshops on water systems and earthquake related aspects. He is a member of four professional organizations including ASCE since 1984, and is a member of the Technical Council on Lifeline Earthquake Engineering seismic risk and water and sewage committees.

Dr.ir. Pieter van Gelder is full-time associate professor of probabilistic methods in civil engineering at Delft University of Technology. He has been involved in research and education on safety and reliability for over 10 years. His research interest is in risk-based hydraulic structural design, and extreme value statistics for hydraulic loads determination. Dr. Van Gelder has worked for the Ministry of Water Management on hydraulic engineering projects where high degrees of reliability were demanded and where statistical information was scarce. Van Gelder is also consultant for Nedstat BV on statistical analysis, and part-time lecturer at IHE, and PAO. He has authored and co-authored over 100 conference - and journal papers, and has supervised over 30 MSc-students from the Faculty of Civil Engineering. He is currently management board member of ESRA (European Safety and Reliability Association), and project leader in several international research projects. Van Gelder is furthermore member in a number of Technical Programme Boards and Scientific Committees of annual international conferences and workshops. Van Gelder also holds visiting professorships at the Institute of Risk Research (IRR) of the Faculty of Civil Engineering of the University of Waterloo, Canada, and at the Centre for Ships and Ocean Structures of the Norwegian University of Science and Technology, Trondheim, Norway.

Stephen Harmsen studied mathematics at University of Washington, Central Washington State College, and the University of Oregon. MS 1973. He has worked in seismological research at the U.S. Geological Survey since 1977. Principal interests are earthquake modeling using finite-difference method, seismic network operations, and PSHA. The latter has been his main focus since 1996, when he joined A. Frankel's group in Golden. He has published in site response, strong-motion attenuation, focal mechanisms and stress field modeling, and seismic hazards of sedimentary basins. Within the PSHA discipline, his main interests are in seismic-source deaggregation, logic-tree analysis, and development of Web-based PSHA tools.

Charles Huyck is a geographer specializing in the integration of advanced geospatial technologies and emergency management. As Senior Vice President of ImageCat, Inc., he oversees a team of engineers, scientists, and programmers developing software tools and data processing algorithms for loss estimation and risk reduction. He has over ten years of experience in GIS analysis and application development. He introduced GIS and Remote Sensing to EQE International, where he served as GIS Programmer Analyst on several loss estimation and research projects. At the California Governor's Office of Emergency services, he was responsible for geospatial analysis, database development and mapping disaster information for the Northridge Earthquake, California winter Storms, and California fire storms. Programs where he has contributed to GIS or remote sensing development include : the HAZUS flood module, EPEDAT, J-EPEDAT, EPEDAT-LA, USQuake, RAMP, U-RAMP, REDARS, Bridge Hunter/Bridge Doctor, VRS, VIEWS, GeoVideo, and a bare earth algorithm for use with SAR and LIDAR data. Research interests include damage detection and inventory development with remotely sensed data. Mr. Huyck has a Bachelor of Science degree in geography from the University of Iowa.

Le Val Lund, P. E., M. ASCE is a Civil Engineer, in Water Resources and Lifeline Earthquake Engineering. He received a BSCE, California Institute of Technology; an MSCE, University of Southern California; a Certificate in Business Administration, University of California at Los Angeles. He is a member and Past Chair, ASCE Technical Council on Lifeline Earthquake Engineering (TCLEE). He is a member and Past Chair, TCLEE Earthquake Investigation Committee and Member, TCLEE Seismic

Risk Committee. He has participated in more than twelve earthquake investigations, including Japan, Turkey and El Salvador. Le Val is a former manager in the planning, design, construction, operation and maintenance divisions for the Los Angeles Water System.

James E. Moore, II, Ph.D. is Professor of Industrial and Systems Engineering; Public Policy and Management; and Civil Engineering, at the University of Southern California. He was born in Newport, Rhode Island, and received his BS degrees in Industrial Engineering and Urban Planning in 1981 from the Technological Institute at Northwestern University in Evanston, Illinois. He received his MS degree in Industrial Engineering from Stanford University in 1982, his Master of Urban and Regional Planning degree from Northwestern in 1983, and his Ph.D. degree in Civil Engineering (Infrastructure Planning and Management) from Stanford in 1986. He specializes in transportation engineering, transportation systems, and other infrastructure systems. He joined Northwestern's Civil Engineering faculty in 1986, and the faculty of the University of Southern California in 1988. He is Director of the Transportation Engineering program, Co-Director of the Construction Management Program, and Chair of the Daniel J. Epstein Department of Industrial and Systems Engineering in the Andrew and Erna Viterbi School of Engineering. This past year, he was elected to the Russian Academy of Natural Sciences, United States Section, for out-standing contributions to the field of Transportation Systems Engineering; and received the Kapitsa Gold Medal of Honor.

Prof. Moore's research interests include risk management of infrastructure networks subject to seismic and other natural hazards; economic impact modeling, transportation network performance and control; large scale computational models of metropolitan land use/transport systems, especially in California; evaluation of new technologies; and infrastructure investment and pricing policies.

Ir. Sonja Ouwerkerk did her master in Civil Engineering at Delft University of Technology. For her thesis, she investigated the impact of river morphology on extreme flood level prediction. A probabilistic approach was used. After her thesis she started doing research on the subject of acceptable risk in collaboration with the Dutch Ministry of Housing, Spatial Planning and the Environment. This research aims at the design of a risk assessment framework with which the government is able to take a rational decision about the acceptability of risks.

Jaewook Park was born in Seoul, Korea. Presently a PhD candidate at the University of Washington, he received his undergraduate degree in Civil Engineering from the University of Minnesota in Minneapolis, Minnesota and his master's degree from Department of Civil and Environmental Engineering at the University of Washington in Seattle, Washington. After the completion of his master's, he joined the structural engineering firm, Skilling Ward Magnusson Barkshire (now Magnusson Klemencic Associates) in Seattle, Washington, where he worked as a structural engineer for three years. Currently, he is completing his doctoral thesis entitled, "An Integrated Approach to Lifeline Performance Evaluation."

Mihail E. Popescu, Ph.D. has more than 30 years of experience in geotechnical engineering research, education and consulting. Main research interests: slope instability investigation and control, engineering behavior of expansive and collapsible soils, soil-structure interaction, computer modeling, geotechnical hazards compilation and assessment. More than 100 research papers in English, French, Italian and Romanian published in refereed journals and conference proceedings. Keynote Speaker and General Reporter at international conferences and symposia. Leader of UNESCO and NATO international research groups and programs. Professor and Visiting Professor at the University of Edinburgh, U. K., University of Tokushima, Japan, University of Natal, Durban, South Africa, Norwegian University of Science and Technology, Trondheim, Norway, University of Civil Engineering, Bucharest, Romania, and Illinois Institute of Technology, Chicago, USA. Responsible for specialist advice to consulting firms, contractors and public authorities on various geotechnical engineering projects. Served the international geotechnical community as Chairman of the Working Group on Landslide Causes of the Geotechnical Societies' UNESCO Working Party on World Landslide Inventory (1990-1995), Co-Chairman of the Commission on Landslide Remediation of the International Union of Geological Sciences Working Group on Landslides (1994-2000), Member of the ISSMGE-IAEG-ISRM Joint Technical Committee on Landslides (2001-present).

Jane Preuss AICP, President of Urban Regional Research, a practicing planner, has two primary specialty areas. One is land use and environmental planning. The other pertains to recovery from and mitigation against the effects of natural hazards, e.g., wind storms, floods, landslides, earthquakes and tsunamis. Aspects of both specialties include damage assessment and underlying geologic and atmospheric causes of damage and public policy context. Aspects of both specialties include integration of engineering research, public policy analysis and regulatory processes. Jane is also active in development of local applications to reduce vulnerability to identified hazards. One example is development of Comprehensive Hazard Mitigation Plans. Jane is active in the Earthquake Engineering Research Institute, and the American Planning Association. She has served on an NAS panel and on several NSF panels.

Dorothy A. Reed, Ph.D. is a professor of civil and environmental engineering at the University of Washington. She joined the faculty in 1983 after a three-year stint at the National Bureau of Standards, now known as The National Institute of Standards and Technology. Her educational background in civil engineering includes a BS from the University of South Carolina and advanced degrees (MSCE and PhD) from Princeton University. Her research has appeared in several publications, including Earthquake Engineering and Structural Dynamics, Journal of Wind Engineering and Industrial Aerodynamics, Journal of Engineering Mechanics and Journal of Structural Engineering. She has lectured extensively in the US and abroad on reliability and structural engineering topics.

Costas Synolakis was born in Athens, Greece, and is a Professor of Civil, Environmental, Mechanical, and Aerospace Engineering at the University of Southern California. He has research interests in tsunami run-up, computer tomography, vibration, isolation of art objects, earthquake hazards reduction, and primate grooming behavior; and has completed $3.6 million in funded research at USC. He completed his BS degree in Engineering and Applied Science, and his MS and PhD degrees in Civil Engineering all at the California Institute of Technology. He has received the National Science Foundation's Presidential Young Investigator Award and the Alexander Onassis Public Benefit Foundation Fellowship. Prof. Synolakis is the author of 49 peer reviewed papers, 28 book chapters of full-length papers, and four books. He has led or co-organized 16 international tsunami surveys, and participated in numerous television documentaries, national and local news appearances worldwide.

Craig Taylor is President, Natural Hazards Management, Inc. and Research Associate Professor, Department of Civil and Environmental Engineering, University of Southern California. He belongs to five professional organizations and has published over sixty papers on risk evaluations and risk policy, with emphases on finance and lifeline systems. For the Federal Emergency Management Agency, he was the project manager on a multi-year project to assess the feasibility of incorporating risk reduction measures into a federal earthquake insurance program, should one be developed. This effort resulted in FEMA-200 and FEMA-201 as well as legislation drafted by Congress. He has been the editor and contributor to TCLEE Monograph 2, *Seismic Loss Estimates for A Hypothetical Water System*. With Elliott Mittler and LeVal Lund, he produced TCLEE Monograph 13, *Overcoming Barriers: Lifeline Seismic Improvement Programs*. With Erik VanMarcke, he edited *Acceptable Risk Processes: Lifelines and Natural* Hazards, TCLEE Monograph No. 21. He is a member of five professional organizations. He received a B.S. in philosophy (mathematics minor) from the University of Utah and an A.M. and Ph.D. from the University of Illinois in philosophy (logical theory).

Erik VanMarcke is Professor of Civil and Environmental Engineering at Princeton University. He was on the faculty of the Massachusetts Institute of Technology until 1985, since receiving his doctorate there in 1970. At MIT, he was the Gilbert W. Winslow Career Development Professor and served as the Director of the Civil Engineering Systems Methodology Group. He held visiting appointments at Harvard University, the Technical University of Delft (Holland) and the University of Leuven (Belgium), his undergraduate alma mater, and was the Shimizu Corporation Visiting Professor at Stanford University. His principal expertise is in engineering risk assessment and applied systems science. He authored *Random Fields: Analysis and Synthesis*, published by the MIT Press, and extended this work to modeling space-time processes and complex systems. He won the Raymond Reese and Walter Huber research prizes of the American Society of Civil Engineers, was awarded a Senior Scientist Fellowship from the Japanese Society for the Promotion of Science, and is a foreign member of the Royal Academy of Arts and Sciences of Belgium.

J. K. Vrijling is a Professor at Delft University. He finished his masters study at Delft University of Technology in 1974. In 1980 he received his masters degree in Economics at the Erasmus University. After a short period at the engineering office of the Adriaan Volker Group he was seconded to the Easternscheldt storm surge barrier project. In this project, he developed the probabilistic approach to the design of the barrier. After the completion of the barrier in 1986, he became deputy head of the Hydraulic engineering branch of the Civil Engineering Division of Rijkswaterstaat. In 1989 he was responsible for the research and computer activities of the Civil Engineering Division. In 1989 he became professor in Hydraulic Engineering in Delft. Since 1995 he has been a full professor in Delft, and advisor to the Civil Engineering Division.

Yumei Wang has expertise is in science, engineering, and technology policy; natural-hazard analyses and risk reduction, and sustainable development. Since 1996, Yumei has been the supervisor of the Geohazards Section at the Oregon Department of Geology and Mineral Industries where she concentrates on earthquake and landslide risks. Yumei served a one-year term as a Congressional Fellow for Senator Ted Kennedy in Washington DC. The fellowship was hosted by the American Association for the Advancement of Science (AAAS) and was funded by the American Society of Civil Engineers (ASCE). Yumei has influenced public policies in her expertise areas as well as in the environment, energy, and transportation in both the state and federal government. She has over three-dozen technical publications, serves on several advisory commissions and committees, and has spoken at numerous conferences. She is an adjunct professor at the Portland State University Civil and Environmental Engineering Department. Before coming to Portland in 1994, she had a geotechnical consulting practice in Oakland, California. She earned her master's degree in Civil Engineering at the University of California in Berkeley in 1988 and her bachelor's degree in Geological Sciences at the University of California at Santa Barbara in 1985.

Manoochehr Zoghi, Ph.D., an Associate Professor in the Department of Civil and Environmental Engineering and Engineering Mechanics at the University of Dayton, joined the faculty in August 1986 after a three-year consulting with the firm Lockwood, Jones, and Beals in Dayton, Ohio. During his tenure at the University of Dayton, he has been teaching and conducting research in his specialty areas of soil-structure interaction, geotechnical earthquake engineering, and infrastructure revitalization. He is the Editor-in-Chief of the International Handbook of FRP composites in Civil Infrastructure (in progress) to be published by CRC-Press. In collaboration with his colleagues at the University of Dayton, he has initiated a study abroad program entitled *Geohazards Norway Natural Hazards and Planning in Europe*, in conjunction with the Norwegian International Center for Geohazards (ICG). His educational background in civil engineering includes a BS and M. Eng. (Master of Engineering) from University of Louisville and a Ph.D. from University of Cincinnati.

APPENDIX B

ROSTERS OF THE TWO PARTICIPATING COMMITTEES AND REVIEWERS OF SUBMITTED PAPERS

MEMBERS OF THE RISK AND VULNERABILITY COMMITTEE, COUNCIL ON DISASTER REDUCTION

Rodrigo Araya
Don Ballantyne
Graham Cook
Jonathan Dollard
Shou Shan Fan
Michael P. Gaus
James Heaney
Antoine Hobeika
Do Y. Kim
Anne Kiremidjian
Frederick Krimgold
Brian Lee
Le Val Lund
Lance Manuel
Kishor C. Methta
Jon A. Pererka
Mihail Popescu
Keith Porter
Adam Rose
Charles Scawthorn
Vijay P. Singh
Nicholas Sitar
John Sturman
Craig Taylor, Chair
Chrisian Unanwa
Erik VanMarcke
Phillip Yen
Manoochehr Zoghi

MEMBERS OF THE SEISMIC RISK COMMITTEE, TECHNICAL COUNCIL ON LIFELINE EARTHQUAKE ENGINEERING

Nesrin I. Basoz
Kenneth W. Campbell

Michael A. Cassaro
Amar Chaker
Chein-Chi Chang
Luke Cheng
Charles B. Crouse
Craig A. Davis
Armen Der Kiureghian
Ronald T. Eguchi
John M. Eidinger
William F. Heubach
David Hu
Jeremy Isenberg
Edward Kavazanjian, Jr.
Mahmoud Khater
Stephanie King
Anne S. Kiremidjian
Shih Chi Liu
Robin K. McGuire
Lamine Mille
Christian P. Mortgat
Dennis K. Ostrom
Keith Porter, Chair
Christopher Rojahn
Dario Rosidi
Jean B. Savy, Vice Chair
Charles Scawthorn
Hope A. Seligson
Masanobu Shinozuka
Craig E. Taylor
Gabriel Toro
Erik H. Van Marcke
Federico Waisman
Leon Ru-Liang Wang
Stuart D. Werner
Masoud M. Zadeh

REVIEWERS OF SELECTED SUBMITTED PAPERS

Kenneth Campbell
Amar Chaker
Chein-Chi Chang
Luke Cheng
Arthur Frankel
Jonathan Godt

Edward Kavazanjian, Jr.
Charles Menun
Dennis Ostrom
David Perkins
Keith Porter
Adam Rose
Masanobu Shinozuka
John Sturman
Robert Shuster
Federico Waismann
Christian Unanwa

Index